The
Holocene

To Glyn and Betty for putting me on the right road,
and to Sylvie for keeping me there.

The Holocene

An Environmental History

SECOND EDITION

Neil Roberts

Blackwell
Publishing

BLACKWELL PUBLISHING
350 Main Street, Malden, MA 02148-5020, USA
108 Cowley Road, Oxford OX4 1JF, UK
550 Swanston Street, Carlton, Victoria 3053, Australia

First published 1989

Second edition published 1998
Reprinted 2000 (twice), 2002, 2004

*Library of Congress Cataloging-in-Publication Data has been
applied for*

ISBN 0–631–18637–9 (hbk)   ISBN 0–631–18638–7 (pbk)

A catalogue record for this title is available from the British
Library.

Set in 10 on 12pt Melior
Printed and bound by Graphicraft Ltd, Hong Kong

The publisher's policy is to use permanent paper from mills
that operate a sustainable forestry policy, and which has been
manufactured from pulp processed using acid-free and elementary
chlorine-free practices. Furthermore, the publisher ensures that
the text paper and cover board used have met acceptable
environmental accreditation standards.

For further information on
Blackwell Publishing, visit our website:
www.blackwellpublishing.com

CONTENTS

TECHNICAL BOXES

PREFACE TO THE SECOND EDITION

Since the late 1980s, when the first edition of this book was written, the theme of Holocene research has proved increasingly topical, perhaps linked to wider concerns over global environmental change. A new journal dedicated to the Holocene started in 1991 edited by John Matthews, and much new empirical information has been 'discovered' in the last decade or so, of which the Wrangel Island dwarf mammoths and the Greenland Summit ice cores are but two of the most spectacular examples. Important new data syntheses have also been published, such as the COHMAP volume (Wright et al., 1993) and the OIES *Global Changes of the Past* (Bradley, 1991), which helped set the agenda for the current PAGES programme. There have been new texts including Chambers (1993) and Bell and Walker (1992). And there have emerged powerful conceptual ideas, such as the role of NADW formation in Sub-Milankovich climate change (Broecker et al., 1990), and methodological progress, for example, in the application of numerical techniques to palaeoecological data (e.g. Birks et al., 1990). All of these demanded to be incorporated in a new edition of the book in order to bring it up to date. I have also taken the opportunity to include some older literature, particularly in the form of additional case studies, both regional (e.g. Mesoamerica, Iceland) and thematic (e.g. disease ecology). The result is that more new bibliographic items have been added to this new edition than there were listed in the whole of the original version. Such is the pace at which Holocene research has been published.

The biggest change from the first edition, however, has been in Holocene chronology. In the mid-1980s the bristlecone pine calibration of the ^{14}C timescale covered about the last 8000 years. Beyond this lay uncertainty; it was not even known if ^{14}C and calendar timescales continued to diverge into the Pleistocene. The logical case for using a calibrated timescale, where possible, is a compelling one, as archaeologists agreed more than two decades ago. But to have to jump from an uncalibrated to a calibrated timescale in the early Holocene, a period of rapid environmental change, was only likely to cause confusion and was therefore not a feasible option at the time when this book was originally written. Instead, the first edition fudged the issue by using uncalibrated dates, which were stitched on to historical chronologies in the late Holocene, when the age discrepancy between the two was small.

By the mid-1990s there existed an extended tree-ring calibration for the Holocene, AMS-dated laminated lake sediments of Late-glacial age, counts of annual snow layers in the GRIP and GISP2 ice cores and U-Th calibration based on corals. These different techniques do not provide identical results, especially for the problematic Holocene–Pleistocene boundary, with its important ^{14}C plateau. None the less, they cross-match sufficiently well to make it possible to use a calibrated timescale for all of the events and changes covered in this book, that is, since about the time of the last glacial maximum. In consequence, I have moved over entirely to a calendar timescale, with ages expressed as Cal. yr BP except for the last 500 years, for which the AD convention is used. Raw ^{14}C dates, as primary data, are normally considered sacrosanct and are left uncalibrated in the research literature. To do so also here would offer two different timescales to the student reader within a single book, and would be a recipe for confusion. So individual dates too have been converted to their calendar equivalents (although a conversion table is included in the appendix at the end of the book for those who will feel the need to refer back to a familar uncalibrated timescale). The change to a calendar timescale will, I suspect, present fewer difficulties for students than for lecturers, for whom a 'gestalt' switch in thinking may be required. I, for one, will be sorry to abandon the elegant simplicity of the round 10 000 years for the start of the Holocene; 11 500 Cal. BP (the replacement date for which I have opted) does not feel half so comfortable!

Note on Book Chronology

Calendar years before present (Cal. yr BP) are used throughout in the revised edition.

ACKNOWLEDGEMENTS

Holocene research is inherently an interdisciplinary venture, drawing its source material from disciplines as diverse as biology, archaeology, geomorphology/geology and climatology. All of these are included within the realm of this book, but the resulting synthesis aims to 'look to the past to interpret the present', and as such is above all a work of geography. Geography seems particularly well fitted to a synthesizing role, with its concern for variations in nature and culture across space and through time. As well as the debt to my home discipline of geography, I am enormously grateful to colleagues who have over the years influenced ideas which appear in this book or who have commented on the text, among them Philip Barker, Richard Battarbee, Morag Bell, Sytze Bottema, Robin Butlin, Denis Cosgrove, Carol David, Oğuz Erol, Andrew Goudie, Dick Grove, David Harris, Gordon Hillman, Roger Jones, John Kutzbach, Frank Oldfield, Bas Payne, Jim Ritchie, Andrew Sherratt, Alayne Street-Perrott, Claudio Vita-Finzi, Tom Webb, John Wilkinson, Herb Wright, and Willem van Zeist. Errors and omissions are of course my own. Thanks are due to the Director and researchers at the CNRS Laboratoire de Géologie du Quaternaire in Marseille, where much of the early writing of the book took place when I was on sabbatical leave. Thanks also go to Val Pheby for secretarial help, Ann Tarver for cartography, Vernon Poulter for photographic assistance, and the Department of Geography at Loughborough University. My family showed great tolerance while I was engaged in writing this tome, and to them I shall be ever grateful. Finally, a special debt is owed to 'la lumière et les paysages du Muy', which supplied the original inspiration for this book.

In producing this new edition I am delighted to acknowledge the help provided to me by others. The extensive redrafting of diagrams was carried out in record time by Peter Robinson, Erica Millwain and Linda Dawes. The Office staff at the Loughborough University Geography Department, and in particular Gwynneth Barnwell, helped with retyping the original text. For comments and assistance I am also most grateful to Jane Reed, Keith Bennett, Charles Turner, Heinkki Seppä and Sytze Bottema. Staff at Blackwell Publishers, past and present, were also instrumental in ensuring that this revised edition finally appeared, and my thanks go to Jill Landeryou, Emma Gotch, Leanda Shrimpton and – above all – John Davey.

The author and publisher would like to thank the following for permission to redraw and reproduce figures: Dr Bent Aaby (6.2); Academic Press (6.4); American Association for the Advancement of Science (4.6, 5.1, 6.5); *Archaeology in Oceania*, University of Sydney (6.9); Dr Keith Bennett (4.5); Professor Jim Bowler (2.13); Cambridge University Press (6.6); Ecological Society of America (4.4, 6.3); Dr John Evans (6.12); Professor John Flenley (3.3); Professor Jack Harlan (5.1); Professor John Kraft (6.5); Macmillan Magazines Ltd. (2.13, 3.2, 4.2, 4.9, 6.2); University of Minnesota Press (2.9); Dr Peter Moore (6.4); National Research Council of Canada (7.9); Professor Frank Oldfield (1.1); The Trustees, *New Phytologist* Trust (4.5); The Prehistoric Society (6.13); *Quaternary Research*, Quaternary Research Center AK-60, University of Washington (4.1); Professor Jim Ritchie (4.9); the Royal Society of London (7.8); Seminar Press (6.12); Dr Richard Tipping (3.4); Dr Jef Vandenberghe (3.6); Dr Paul Waton (6.15); John Wiley & Sons (4.3, 7.6); Professor Willem van Zeist (5.7, 6.11); Dr Kathy Willis (4.4).

The author and publisher are also grateful to the following for permission to use and for their help in supplying photographs: Swissair (2.6), British Antarctic Survey (photo: C. Swithenbank) (3.1), *Philosophical Transactions of the Royal Society of London* (B.265, pp. 298–326), R.B. Angus, 1973 (3.2), Novosti Press Agency (photo: A. Zubtsov) (3.4), Professor David E. Smith, Coventry University (4.1), Dr Nicole Petit-Maire (4.4), Sonia Halliday Photographs, photo James Wellard (4.5), Robert S. Peabody Foundation for Archaeology (5.3), Robert Harding Associates (5.4), Oxford Scientific Films, photo Hjalmar R. Bardarson (6.1), Dr Jon Pilcher, Queen's University, Belfast (6.2), Museum of New Zealand Te Papa Tongarewa, C.73 (Photo: Augustus Hamilton) (6.3), Hulton-Getty (6.4), University of Cambridge, Committee for Aerial Photography (6.8), Corbis/Bettmann (7.1), Dr Timothy Bayliss-Smith (7.4, 7.5), Southwark Local Studies Library (7.3), Dr André Lotter (7.4), Dr Carl Sayer (7.6). All other photographs are by the author.

CHAPTER ONE

Introduction

Are the moors and downs of the British Isles natural landscapes or were they created by human agency? To what extent has soil loss through erosion increased on cultivated land compared with areas of natural woodland? And is the recorded rise in global temperatures since 1980 a response to an increase in greenhouse gases in the atmosphere, or just part of the climate's natural tendency to vary through time? Answers to problems such as these are often sought by monitoring contemporary environmental processes. For example, Gordon Wolman (1967) measured sediment yields from forested and cultivated land in the eastern United States, and found that erosion was up to eight times higher for the latter, while Anders Rapp and others (1972) recorded over a hundredfold increase from erosion plots in Tanzania. Alternatively, data from gauged field stations may be used, although the short time period of observation often proves to be a handicap. Meteorological stations, for example, normally have records only spanning the twentieth century (Hulme, 1994), and this makes it hard to identify accurately any long-term warming trend.

In fact, neither monitoring nor gauged records are likely on their own to provide complete solutions to the problems posed. This is because a longer-term view is required. The 'three score years and ten' of an average human lifespan is so much shorter than the millennia of natural history that we tend to be aware only of short-term variations in the environment – the wet summer, the late spring, the 'record' flood. What we are much less aware of are slower, subtler changes such as alterations in the floristic composition of woodland, the silting up of estuaries, and the advance and retreat of glaciers. Yet only a long-term perspective could tell us that Britain's downs and moors were transformed from their original woodland before written history even began.

Sources of Information on Past Environments

For historic times documentary sources can sometimes provide reliable observations on the former state of the natural environment. Tax records have been used to indicate late seventeenth- and early eighteenth-century climatic deterioration in southern Norway (Grove and Battagel, 1981), while maps and legal documents show the changing position of Spurn Point spit on the east coast of England during the last 300 years (de Boer, 1964). Among the best historical sources are the early United States federal land surveys, which mapped the pre-existing forest and grassland vegetation, and even recorded their species composition, before laying out land boundaries. Documentary data of this sort form an important part of historical ecology which applied to past flora and fauna (Sheail, 1980), historical geology when related to changes in the physical landscape such as rivers (Hooke and Kain, 1982), and palaeoclimatology when linked to former climates (Bradley and Jones, 1992). In practice, the distinction between these three sub-disciplines is often an arbitrary one.

Written records are, however, restricted to literate cultures, and as late as AD 1500 these existed only in Europe, Asia and North Africa. There are consequently long periods of human history for which recorded observations are absent, termed by archaeologists Prehistory. Moreover, written history covers very different timespans in different parts of the world. Whereas written history began around 5000 years ago in Mesopotamia with the Sumerians, and 2000 years ago in Britain with Julius Caesar, it only started in the 1930s in the highlands of Papua New Guinea when aircraft brought the first European contact. Prehistory has come to be associated in the European mind with all that is remote, both in time and in cultural affinity, to twentieth-century life. On the other hand, to a New Guinea highlander (or a Maori or black Zimbabwean) Prehistory re-presents a direct cultural heritage which ended only a few generations ago. All of this means that the attitudes towards the natural world of many past societies have either gone un-recorded, or have appeared in written form only through the eyes of others. Ethnobotanical and other studies are, however, beginning to reveal a remarkable indigenous knowledge of the natural environment and its uses by modern non-literate hunter-gatherers, peasant farmers and nomads (Myers, 1979; Richards, 1985).

It is not only from ancient times and distant places that historical data on the natural environment prove to be defi-cient. Old men may recall how in their youth fine catches of salmon were taken from Scandinavian rivers now devoid of fish, but because no one recorded the pH of the water until acid rain became a problem during the 1970s, it is difficult to know whether the salmon were eliminated by **acidification** or simply overfished. Availability of documentary sources there-fore tends to inhibit consideration of non-literate regions, such as Papua New Guinea, and problems such as acidification, which do not appear in historical accounts. The example of megafaunal extinctions (see chapter 3) serves to illustrate the point. When told that much African wildlife is threatened by extinction, our reaction is one of horror. Yet extinctions of even greater magnitude have occurred within the timespan of human history considered in this book, notably in the Ameri-cas where 90 per cent of all large mammals were lost. How-ever, this ecological crisis remains known only to a narrow audience of specialists because no contemporary written accounts of it were handed down to posterity.

This book aims to recount and help to explain changes in the natural world through time, including those in human–environment relations. If a long-term and cross-cultural perspective is to be taken on this, then it is necessary to escape from the restrictions imposed by information derived solely from contemporary or documented historical sources. This is not to deny the critical importance of such sources, but while they have been employed extensively in studies of environmental relations (e.g. Glacken, 1967; Pepper, 1996), other sources have not. Without a time machine in which to return to the past, we have instead to rely on proxy evidence, including that from **palaeoecology**, which provides ecological histories, and archaeology, which provides cultural histories. These two sources sometimes come together as **bio-archaeology**, in which plant and animal remains from archaeological sites are studied to reconstruct their economy and environment. Additional data come from historical geology and **geo-archaeology**. These subjects share in common many techniques such as **radiocarbon dating** and pollen analysis, and they are often investigated together to form interdisciplinary research projects.

For the moment it is sufficient to note some of the strengths and weaknesses of the palaeoenvironmental approach (Rymer, 1980). One of its weaknesses is that it cannot reconstruct attitudes to the natural world by former human societies in the way that historical sources can. Medieval cosmologies which placed humanity in a holistic relationship with the natural world would, for example, be scarcely comprehensible without textual explanation (Pepper, 1996). Bio-archaeology, with its concern for site economies and food remains, inclines the investigator towards an economic view of past human life (e.g. Higgs, 1975). This bias towards economic and away from social and cultural explanations is most easily countered where the archaeological past meets the anthropological present, for instance with the Hopi Indians of the American Southwest (Butzer, 1982). In the case of prehistoric Europe it is not so easy.

On the other hand, one of the great advantages of archaeology and palaeoecology is their ability to identify long-term patterns of cultural or environmental development. This is especially important because the timescales of adjustment in many environmental systems span centuries or millennia rather than individual years, via processes such as soil maturation and ecological succession (see figure 1.1). In the case of palaeoecology, environmental reconstruction can be applied to recent as well as long-term changes. Most lake sediments, for instance, have continued to accumulate up to the present

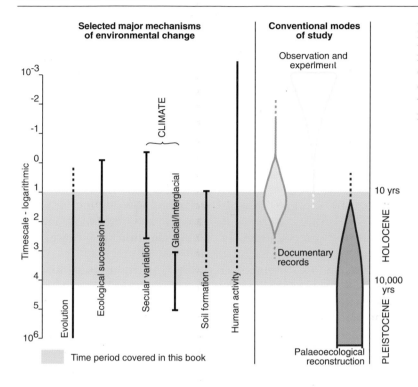

Figure 1.1
Mechanisms and
modes of studying
enviromental change
over different
timescales (modified
from Oldfield, 1983)

day, and they can therefore be used to give an historical dimension to such contemporary environmental problems as acidification of freshwaters. The continuous nature of most palaeoecological records provides what Frank Oldfield (1977) has aptly described as 'a true continuum of insight'. This enables us to establish whether ecological changes have been episodic or gradual, and to identify base-line conditions if they existed. These techniques, unlike written records, have the further advantage of being as applicable in New Guinea as they are in New England.

Nature and Society

All the world's landscapes and ecosystems are products of the natural and cultural processes that have shaped them over time to bring them to their present state. But what have been those processes? Over the last 1 million years a major factor has been climatic change, with the climate oscillating between **Pleistocene** glacials and interglacials. Only after the end of the last glaciation, c.10 000 years ago, did the world's climates and environments take on a recognizably modern form. Secular climatic variations of smaller magnitude and shorter duration have continued up to the present day. But whereas environmental changes brought about by natural agencies such as climate essentially diminish in amplitude as one moves forward in time, there is one set of processes which has done precisely the reverse; human impact on the environment has increased progressively through time as *Homo sapiens* has been transformed from hunter to city-dweller.

The oldest and simplest human mode of production is hunting, fishing and gathering (h-f-g), described by Mandel (1969, p. 34) as 'a primitive economy . . . in which the results are so meager that they must be shared to avoid death by starvation'. In fact, a wide range of anthropological studies suggests that far from eking out a meagre existence in a hostile environment, h-f-g groups probably represent the original affluent society (Sahlins, 1974). Old World h-f-g populations are represented archaeologically by **Palaeolithic** and **Mesolithic** cultures.

Hunting and gathering was followed in the process of social evolution by peasant farming, the two being distinguished by the latter's exploitation of domestic plants and animals. The adoption of agriculture by h-f-g groups, termed the **Neolithic** revolution by Vere Gordon Childe, may have been the most decisive revolution in the whole of human history (Smith, 1995). It undeniably changed the basis of human relations with nature, and probably explains why history subsequently followed different courses for different peoples (Diamond, 1997). The basic form of production in h-f-g and subsistence agricultural economies is simple and communal, with no systematic expropriation of surplus labour. This is not true, however, of more complex agricultural economies which supported feudal, classical and 'hydraulic' societies. The resulting states and civilizations typically – although not always – brought literacy and the start of written history. Finally, the last two centuries of human history have witnessed the rise of industrial, and arguably post-industrial, economies and the global expansion of western European culture. Each stage in this social evolution of humankind has seen an increase in control over our relationship with nature (Simmons, 1996). As we have evolved from hunting to agriculture to industry and beyond, so human impact upon the environment has apparently come to counterbalance or even replace environmental influence over human affairs.

A great strength of this evolutionary approach to human–environmental relations is that it is historically mediated. However, consideration of the historical dynamic is usually restricted to the human half of the partnership. The forces of nature are all too often viewed as an essentially passive backcloth against which human history is acted out. Although this environmental backcloth may change from one scene to another, it plays little or no active part in its own right. Perhaps the sky will be permitted to vary to show storm, snow or sun as a token gesture to the fact that nature is not static, but the forests and streams will be ever present, the natural landscape constant. It is above all this ahistorical view of nature that this book seeks to challenge. Climate, forests and rivers have their histories, too.

The Significance of the Holocene

The period over which most of these cultural and environmental changes have taken place is the **Holocene**. The Holocene, or post-glacial epoch, provides the time frame for this book, which is organized in chronological fashion, starting with the last glacial-to-interglacial transition and working towards the present.

During the late Pleistocene, from about 25 000 to a little over 10 000 years ago, the glacial climate made the Earth cold and unfamiliar. Canada lay buried beneath several km of ice, Europe was devoid of forests, and the southeast Asian islands were joined to form a single land mass – these are changes described in chapter 3. The climatic shift at the end of the last glaciation was important not only in its own right, but also because it indirectly controlled other elements of the environment. Post-glacial adjustments of plant and animal distributions, sea levels, geologic and soil-forming processes, are discussed in chapter 4. The first half of the Holocene would be the period to look to if we were to search for nature's primaeval, virginal base-line state; one free from 'significant' human disturbance. However, this condition was to end with the **domestication** of plants and animals. The emergence of farming and its initial impact upon European woodland are considered in chapter 5. Chapter 6 focuses on later Holocene environments, notably those produced by complex agro-ecosystems such as developed around the Mediterranean basin. Finally, chapter 7 discusses the historical impact of industrial society upon pollution, and of European expansion overseas upon land use and ecology.

But before embarking on this natural history of the last ten thousand years, we need to establish how it is to be reconstructed. For this reason the next chapter of this book is devoted to describing some of the main techniques of Holocene environmental reconstruction.

CHAPTER TWO

Reconstructing Holocene Environments

Dating the Past

Dating techniques are fundamental to an understanding of the natural and cultural changes which have taken place during the Holocene. Without them events such as the Neolithic (agricultural) revolution would float aimlessly in time. Perhaps worse, without independent dating methods we would be forced to depend on environmental and archaeological evidence to provide chronologies, thus robbing this evidence of much of its potential meaning (Vita-Finzi, 1973).

Curiously, the most obvious reason for wanting to date the past may be the least important; that is, the desire to know how old something is simply for its own sake. It may stagger the imagination to think that there are trees living today in California's Sierra Nevada which were saplings before Stonehenge was completed, but the significance of neither henge monuments nor the bristlecone pine is better explained as a result. A much more significant role for independent dating lies in the testing of hypotheses (Deevey, 1969). Take, for example, the mid-Holocene elm decline widely recorded in pollen diagrams from northwest Europe. If, as has been hypothesized, this was a consequence of prehistoric agriculture, then the fall in elm pollen values should coincide with the time of arrival of the first farming communities. Dating early Neolithic sites, on the one hand, and the appropriate section of pollen diagrams, on the other, allows the anthropogenic hypothesis to be tested. In particular, if the elm decline is found to pre-date the start of the Neolithic, then this hypothesis would be shown to be false, and our agricultural ancestors exonerated from blame (see chapter 5 for further discussion of the elm decline).

Another reason for wanting to have precise estimates for the age of Holocene events is in order to calculate past rates of change. Some events, such as **eustatically** controlled sea-level rise, would have occurred at the same time everywhere across the ocean; that is, the change was a **synchronous** one. More usually, however, events begin earlier at some places than at others. The spread of a disease or the movement of a glacier snout are of this type, and they are termed **time-transgressive**. The rate of disease diffusion or of glacier retreat will vary between cases and precise dating techniques make it possible to establish whether rates were fast or slow, constant or variable. Dating is also essential to a number of the other techniques discussed later in this chapter. Past influx of pollen or of sediment into a lake, or of algal productivity within it, can only be obtained if a sound and detailed chronology exists for the sequence of lake sediments.

One simple form of dating is provided by the fact that in undisturbed sediments younger layers overlie older ones. This law of superposition indicates which layer was deposited first, but it fails to provide the actual age of either. Preferably, the layers should be placed not only in a relative sequence such as this but also firmly in time, and for this it is necessary to assign them absolute ages in years. Four principal approaches to absolute dating exist: (1) those based on historical records; (2) those employing **radiometric dating** techniques; (3) those utilizing incremental dating methods; and (4) those based on palaeomagnetism. Each will be discussed here, but most attention will be paid to the second and third approaches, which will be illustrated in detail by reference to the specific techniques of radiocarbon dating and **dendrochronology**.

Historical dating

This form of dating is most obviously associated with documentary-based studies of historical ecology or climate. For instance, medieval manuscripts recording the extent and condition of England's royal forests almost invariably have dates attached to them, as do Icelandic chronicles referring to drift ice – and hence sea temperature – around that island. We may go further and suggest that without a date, documentary records such as these are of little use in helping to piece together past environments.

Historical or archaeological evidence can also be important in dating non-documentary records such as pollen diagrams. In the case of one core from southwest Syria, the presence of an exotic pollen type, that of maize (*Zea mays*), helped show that the upper part of the core dated to recent centuries and not to the early Holocene as had previously been proposed (Bottema, 1977). Maize is native to the Americas and was introduced into the Old World as a crop only after the Spanish conquest of Mexico in the sixteenth century.

The remnants of past human activity can provide other important clues for dating Holocene environmental changes. A Roman tile drain buried beneath 5 m of alluvium, or a classical port now silted up and many km from the sea, both testify to active sediment transport and deposition by rivers during the last 2000 years. Artifacts, including pottery, stone tools and coins, can all be assigned ages with greater or lesser precision (see plate 2.1), as can less obvious cultural evidence such as hemp fibre from retting that has been incorporated in lake sediments. If discovered in a stratigraphic sequence, artifacts provide maximum ages for the layer in which they were found; maximum because they may have been re-worked since being deposited initially. A bicycle frame pulled from supposedly

Plate 2.1 This Arab coin from Crete, made between 847 and 861 AD, helped date the alluvial fill in which it was found

mid-Holocene coastal dunes provides a – revised – maximum age for the dunes, for even if the bicycle were manufactured 50 years ago it could have been dumped there as recently as last year. On the other hand, structures such as a former irrigation channel or a shell midden will not have been redeposited in the same way as artifacts and they may therefore offer tighter dating control. In fact, built structures will often provide minimum rather than maximum ages, say for a land surface on which they are found. However, the use of human artifacts for dating purposes can be fraught with dangers, as it necessarily involves time-transgressive phenomena. The stone age may have ended 5000 years ago in eastern Europe, but it arguably lives on with certain isolated hunter-gatherer groups in the tropics.

Radiometric dating methods

Radiometric dating techniques have been discovered and applied only during the second half of the twentieth century. They involve the radioactive properties of different materials which contain within them a natural time signal, most often involving the principle of isotopic decay. Most natural elements are a mixture of several **isotopes**, which have the same chemical properties and atomic numbers but different numbers of neutrons and hence different atomic masses. One isotope is always dominant for each element; for instance, in the case of carbon the dominant isotope is carbon-12 (or ^{12}C). Carbon isotopes with atomic masses different from this are ^{13}C and ^{14}C, of which the latter is by far the least abundant.

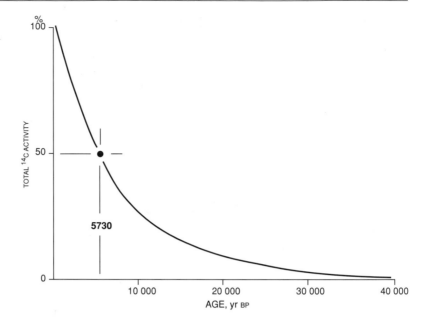

Figure 2.1 Decay
curve for radiocarbon

Carbon-14 is also different from the other two in that it is
isotopically unstable; in other words, it decays to form stable
^{14}N over time. Most importantly of all, unstable isotopes such
as ^{14}C decay at a fixed rate. The rate of radioactive decay is
not a straight line but occurs rapidly to begin with and slows
down progressively over time (see figure 2.1). Because decay
rates are measurable, unstable isotopes represent natural
archaeological or geological clocks. The clock is set to zero
when isotopic decay starts, and the time elapsed can be gauged
from how far the decay process has proceeded since then, as
indicated by the amount of radioactivity left in the element.
The decay rate for a particular isotope is most usually recorded
by its half-life, or the time taken for its radioactivity to be
reduced by half. Half-lives of elements vary from the very short
(e.g. radon-222, 3.8 days) to the very long (e.g. potassium-40,
1300 million years). In general the useful dating range of
individual isotopic methods is about ten times their half-life.
Beyond this, radioactive emissions are difficult to distinguish
from normal background levels of radioactivity. Consequently,
different isotopes provide dating methods for very different
time periods, and only a few of these will be useful for the
Holocene.

The most important of the radiometric dating methods for the
Holocene is based on the isotope carbon-14. RADIOCARBON
DATING was pioneered by Willard Libby at Chicago in the late
1940s, work which later earned him the Nobel Prize for Chemistry (Burleigh, 1981). ^{14}C is formed in the upper atmosphere
by cosmic ray bombardment of nitrogen (N) atoms. The result-

ing carbon isotope is rapidly oxidized to form carbon dioxide (CO_2), and is then mixed uniformally and rapidly through the atmosphere. Photosynthesis leads to ^{14}C being taken up by plants, which in turn is passed on to higher organisms including humans. Consequently, ^{14}C is present in small but significant quantities in all living organisms. When organisms die their ^{14}C content is no longer replenished and the isotopic clock is set in motion. Libby calculated the half-life of ^{14}C experimentally as 5560 ± 30 years, a value close to the present best estimate of 5730 ± 40 years. Libby cross-checked samples which he had dated by the ^{14}C method against materials of known historical age, most of which came from ancient Egypt and were up to 5000 years old. The close level of agreement led him to believe that his radiocarbon dating method was a valid one and that it could be applied universally.

Since its discovery about 100 000 ^{14}C determinations have been carried out by over 100 different laboratories. About half of all dates have been for archaeological investigations, with most of the rest having been for studies of past environments, and with the vast majority of radiocarbon dates being of Holocene age. The results of radiocarbon dating are usually presented as ages in years before present, normally written using the notation yr BP. In practice, the present is taken as AD 1950 in order to prevent dates appearing to be older simply because they were analysed more recently. Archaeological chronologies use the BC/AD system more often than the yr BP one, but the former can be turned into the latter simply by adding 1950 or – as a quick approximation – 2000 years to BC dates.

Also listed along with each individual ^{14}C date is a laboratory code number and an error function representing one standard deviation about the mean. Thus 8160 ± 110 yr BP (BM – 1666) was dated at the British Museum radiocarbon laboratory and indicates that there is a 68 per cent probability that the ^{14}C age of this sample lies between 7940 and 8380 yr BP. It should be borne in mind that radiometrically determined dates are statistical estimates, and they should therefore be treated as such. ^{14}C dating can be used to determine the age of any material containing carbon, including wood, charcoal, peat, seeds, bone, carbonate, shell, iron, cloth, rope, groundwater and soil. Among the more exotic materials dated have been ostrich eggshell, lime mortar and human brain!

The ^{14}C method cannot, however, be applied to samples of very recent age. The burning of fossil fuels since the Industrial Revolution has caused an injection of geologically old carbon into the atmosphere which has lowered its ^{14}C content and made recent samples seem older than they actually are – the so-called Suess effect. More recently still, this has been overcompensated and further complicated by the testing of atomic weapons. In short, ^{14}C dating is of little use for samples younger than about 150 years.

Many other forms of sample contamination are possible, and although ash from a careless investigator's cigarette has doubtless accounted for more than one erroneous ^{14}C date, contamination is usually of natural rather than of human origin. Carbon both older and younger than the date of the death of the sample may be involved. The latter is a particular problem in areas of limestone or coal-bearing bedrock. This problem is especially difficult to assess where bicarbonate-rich lake waters have been taken up by aquatic plants which subsequently formed part of the lake sediment. The dilution of ^{14}C levels causes ages to appear older than they actually are, and this is known as the hard-water effect (Deevey et al., 1954). Contamination by young carbon, by contrast, makes dates appear too recent. There are countless such sources, including rootlets penetrating into underlying stratigraphic layers, recalcification of mollusc shells, and animal burrows in archaeological sites. However, many problems of contamination by younger material can be avoided by careful selection of samples in the field and by pre-treatment during dating. Young carbon is, in any case, a much more serious problem for samples of Pleistocene than of Holocene age. Improved dating precision is also being achieved by a method developed in the 1980s of measuring ^{14}C directly using a mass spectrometer. This **AMS (or Accelerator Mass Spectrometer) dating** method allows the dating of small samples weighing as little as 5 mg. It allows scientists to pick out and date individual seed remains, especially of land plants which should be free of any hard-water error, or charcoal fragments. More controversially, it allowed the age of threads from the Turin Shroud to be determined, which could otherwise have been dated only by destroying the shroud itself. The result showed the Shroud to be a – very clever – medieval forgery, fabricated between AD 1260 and 1390.

Some materials are often considered to be more prone to contamination than others – for example, bone, shell and soil – but ^{14}C dating will be more reliable if samples of these materials are adequately prepared before age determination. In the case of bone, only the protein collagen should be dated (Gillespie, 1984, p. 13), while mollusc shell should be tested by x-ray diffraction and acid leached or mechanically cleaned if necessary (Vita-Finzi and Roberts, 1984). Soil, of course, takes longer to form than either of these two, but the consequent broad time range associated with ^{14}C dates on buried soils is as much a stratigraphic as a dating problem (it also applies, for instance, to the interpretation of soil pollen) (Matthews, 1985). ^{14}C age determination of soil should ideally involve dating selected organic fractions; but even without this, Rothlisberger and Schneebeli (1979) obtained consistent and meaningful results in their study of Holocene soil and moraine stratigraphy in the Swiss Alps. All of these contamination problems are local to sites under study. By contrast, the one major revision to the ^{14}C timescale since Libby's initial discovery involves variations that were global in scale.

It was initially assumed by Libby that there had been no significant changes in atmospheric levels of ^{14}C during recent millennia. Libby had indeed carried out ^{14}C determinations on historically dated materials to check this, but after initial enthusiasm he subsequently found that older dates systematically underestimated the true age of samples. This was confirmed in 1965 when Hans Suess presented results comparing ^{14}C dates with those from another dating method – dendrochronology. These results are discussed in detail below, but suffice it to say that ^{14}C dates older than 2500 years significantly underestimate actual age. But because these deviations from true age are systematic and world wide, it is possible to apply correction, or calibration, factors to ^{14}C determinations to turn them into true, or calendar dates. The need for ^{14}C calibration has not, therefore, invalidated the ^{14}C method, which remains a remarkably robust and successful one. The radiocarbon method and its applications are discussed in more detail by Gillespie (1984) and Olsson (1986), and the AMS method by Hedges (1991).

Two other radiometric techniques of increasing importance are luminescence and uranium-thorium dating. LUMINESCENCE, or Optical, dating was initially applied to archaeological materials such as pottery, burnt clay and flint (Aitken, 1990). It was then discovered that 'bleaching' of the sample, which resets the radiometric clock, could be accomplished by exposure to sunlight as well as by firing. Thermo-luminescence or TL dating subsequently came to be successfully applied to the quartz or feldspar grains of wind-blown sediments such as sand dunes and loess (Wintle and Huntley, 1982), assuming that they were not deposited in night-time darkness! The technique has since added other variants, such as Optically Stimulated Luminescence (OSL), which require shorter 'bleaching' times, and are capable of dating other types of sediment such as river alluvium (Aitken, 1994; Duller, 1996). URANIUM-THORIUM, or U-Th, dating is based on a more complex decay chain than ^{14}C, with a sequence of 'daughter' isotopes being produced from the original 'parent' (Smart, 1991; Ivanovich and Harmon, 1995). On the other hand, the half-lives involved are longer than with ^{14}C, so that the technique can be used to date older materials. Unstable uranium or thorium isotopes are taken out of water by corals, or precipitated in cave speleothems or in lake sediments. U-Th, especially with the use of a high-precision mass spectrometer, has proved valuable in calibrating the radiocarbon method beyond the range of dendrochronology (Bard et al., 1990).

Luminescence, uranium-thorium and ^{14}C all cover timespans of thousands of years or longer, and none is useful for dating changes during the last century. Partly because of this, palaeo-ecological studies tended to stop precisely at that point in time when human impact has often become most apparent. However, a number of dating techniques with short half-lives have been developed which allow the timing of recent human impact on ecosystems to be determined. Probably the most useful of these new radiometric dating techniques are LEAD-210 and CAESIUM-137 (see Technical Box X, pp. 236–8).

Dendrochronology and radiocarbon calibration

Incremental techniques, which are based on natural seasonal rhythms or annual growth rates, have been utilized since the early years of the twentieth century. However, it would be wrong to imagine that they were made redundant by the advent of radiometric dating techniques. As will be made clear from the discussion that follows, they remain as valuable to-day as they ever were.

The principles of this type of dating can be illustrated by what is probably the best-known incremental dating method: tree-ring dating or DENDROCHRONOLOGY. This technique was largely developed through the efforts of A.E. Douglass working in Arizona between 1910 and 1940. Although it had been known since at least the time of Aristotle that trees produced annual growth rings, Douglass was the first to systematize this into a dating method. At its simplest the technique involves counting the number of growth rings present between the bark and the centre of a living tree. As the outermost ring can be assumed to represent the current year's growth, the age of the tree can be calculated by counting the number of rings present. This need not require lumberjack skills, however, for a tree-ring sequence can be obtained without harming the tree by using an increment borer, which provides pencil-sized tree cores. This basic form of tree-ring dating can be used to provide a chronology for valley glacier retreat or alluvial fan activity over recent decades and centuries, although the ages it provides in these cases are only minimum ones because of the time lag between deglaciation or channel change and subsequent colonization by trees.

Potentially of greater value are those approaches which exploit the characteristic tree-ring signatures produced as a result of differential growth from one year to the next. Favourable environmental conditions lead to the formation of a broad growth ring, while adverse conditions have the opposite effect. The result is a unique sequence of wide and narrow rings not unlike the bar-codes used for store price tags. The main environmental factors which cause ring-width variations are climatic ones, but the critical stress factor will not be the same everywhere. In the Alps a narrow tree-ring is usually the product of a cold summer, while in the American Southwest low rainfall is more likely to be responsible for stunted growth. Tree-rings can therefore provide proxy climatic data, the study of which is known as dendroclimatology (Schweingruber, 1988; Fritz, 1991).

Tree-ring signatures also fulfil another function, that of allowing cross-dating or correlation between tree-ring sequences. Such cross-dating is not restricted to live wood, so that sequences from dead trees or wood used in buildings can be matched up with those from living trees (see figure 2.2). The former would otherwise be free-floating chronologically, as their outer edges do not date to the present day and consequently lack a firm reference point in time. The building up of a chronology from separate tree-ring sequences amounts to no less than a jigsaw puzzle with time. 'The pieces are scattered round as living trees, stumps, timbers in buildings and buried, either as archaeological material or as naturally preserved timbers, in bogs, rivers or lake beds' (Baillie, 1995). Douglass, in Arizona, compiled both a master chronology stretching from living yellow (ponderosa) pines back to the thirteenth century AD, and a floating one based on pine timbers from prehistoric buildings. In 1929, after much searching, he found the missing piece in his jigsaw – an excavated wooden beam from Pueblo Bonito dating to between AD 1237 and 1380 – and the two sequences were joined.

Huber subsequently applied Douglass's approach to German tree-rings and by 1963 had completed a 1000-year chronology based on oak rather than pine timbers. This oak chronology has since been extended and provisionally joined to a 1605-year floating pine sequence which together go back right to the beginning of the Holocene (Kromer and Becker, 1993). A similar 7000+ year chronology has been built up from Irish bog oaks by a research team at Queen's University, Belfast, using the same cross-dating approach (Pilcher et al., 1984). The most famous dendrochronological record, however, comes from the bristlecone pine (*Pinus longaeva*) (plate 2.2), which grows at altitudes of 3000 m in the mountains of eastern California and Nevada. This is the world's longest-living tree and, incredibly, living bristlecone pines have been recorded up to 4600 years old (Ferguson, 1968). Cross-dating of sub-fossil trunks with living trees has produced a continuous bristlecone pine tree-ring chronology 8200 years long. The construction of tree-ring

Figure 2.2 Cross-
dating alpine timbers
from overlapping tree-
ring signatures

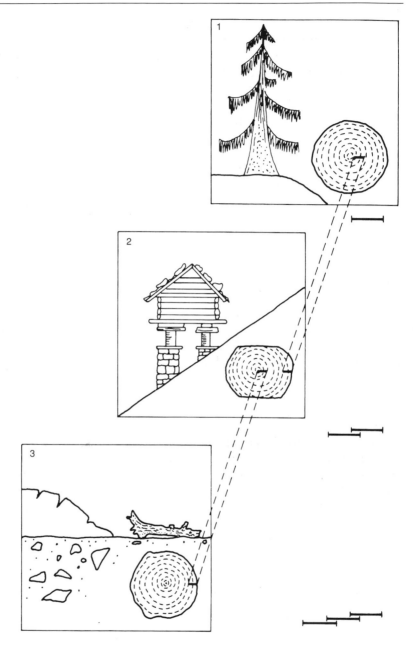

chronologies such as these involves the science of computer
matching and x-ray densitometry as well as the art of matching tree-ring series by eye (Baillie, 1995).

Possibly the most important feature of dendrochronology is
the fact that the wood used for tree-ring counts can also be
dated by radiocarbon, and this has allowed a cross-check to be
made between the two methods. At first sight it may seem
surprising that it is possible to get 'old' dates from 'living'

wood, but most of the woody cells in trees are empty, being used only for transporting water from the roots to the branches. The cells laid down in each growth ring are the only part to retain a new ^{14}C imprint. In the 1960s, Hans Suess dated samples of bristlecone pine of known tree-ring age by radiocarbon. His results revealed a systematic divergence between ^{14}C and tree-ring ages in samples more than about 2500 years old (see figure 2.3). Mid-Holocene ^{14}C dates appear to be too young, with a discrepancy amounting to 600 years (or 12 per cent) for material 5000 years old. Tree-ring dates are more accurate than their ^{14}C equivalents, although this does not invalidate the ^{14}C method, but rather requires that ^{14}C dates be calibrated if they are to be expressed as true, or calendar years. Bristlecone pine calibration caused important revisions to be made to archaeological chronologies based on ^{14}C, and in consequence the theories of culture change in European Prehistory (Renfrew, 1973). Palaeoecologists and other environmental scientists have been more reluctant than archaeologists to convert their ^{14}C dates into calendar years. This is partly because until recently, calibration curves covered only part, not all, of the Holocene. Now, however, tree-ring calibration has been taken further back in time and has been joined by other dating methods which extend to the period of the last glaciation. Two of these, tree-rings and U-Th ages from corals, have been combined in a widely used ^{14}C age calibration computer program, CALIB 3.0 (Stuiver and Reimer, 1993).

There now seems every reason, therefore, to adopt a calendar rather than a conventional, uncalibrated ^{14}C timescale for an environmental history of the Holocene. In consequence, **the ages in this book will use the convention of quoting dates calibrated, expressed as Cal. yr BP** (Calendar years before present). One advantage of this is that it removes the discrepancy between radiometric and historical chronologies during the period from about 5000 to 2500 years ago. Calibration has been based on CALIB 3.0 and OxCal (Ramsay, 1995), with minor modification at the Pleistocene–Holocene boundary (see Technical Box I). A summary conversion table from ^{14}C to calendar ages (and vice versa) can be found in the appendix on p. 253. Calibration has been applied to individual ^{14}C dates as well as to overall chronologies. In most published papers, ^{14}C determinations are quoted only as raw, uncalibrated dates, and they would need to be converted to calendar ages in order to make a comparison with the timescale used here.

It is important to recognize that there are some potential pitfalls in making a conversion from ^{14}C to calendar years. First, the tree-ring calibration curve is not a smooth line but contains a number of wiggles which represent short-term variations in atmospheric ^{14}C content (figure 2.3). This is significant

Plate 2.2 The world's longest-lived tree, the bristlecone pine (*Pinus longaeva*), which lives in the high mountains of the Sierra Nevada in western North America

in that some ^{14}C dates have more than one calendar age; a ^{14}C date of 4500 yr BP for instance has calibrated ages between 5056 and 5246 years. Calibration can therefore have an important effect on studies where high dating resolution is required because an apparently synchronous event may be turned into one that is potentially time-transgressive. This is a particular problem for dating the onset of the Holocene (Technical Box I). In high-resolution studies, on the other hand, it may be possible to obtain sufficient ^{14}C dates to use the 'wiggles' of the calibration curve to match the series being dated. A second

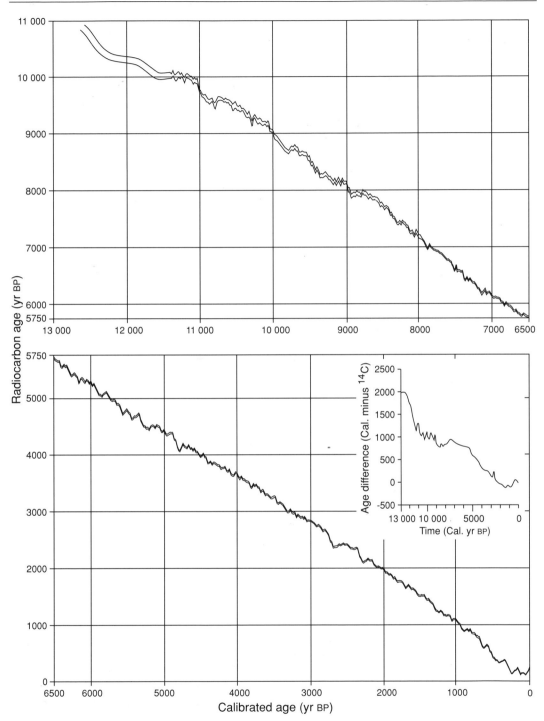

Figure 2.3 Tree-ring radiocarbon calibration curve for the Holocene (inset: departures from calendar age with time) (from OxCal. 14L program; Younger Dryas based on Goslar et al., 1995)

TECHNICAL BOX

Technical Box I: Dating the onset of the Holocene

There are different schools of thought about how the Holocene should be formally defined. Because the Holocene still 'lives', debate has focused on defining its onset, namely the Pleistocene–Holocene boundary. The first school of thought (e.g. Watson and Wright, 1980) believes that the beginning of the Holocene is easily recognized in individual records of deglaciation or biotic change (e.g. in pollen diagrams), but that because these changes are not exactly the same age everywhere, the Pleistocene–Holocene boundary should therefore be time-transgressive. A second school prefers the 'type-section' approach commonly used in geology to define stratigraphic boundaries. Mörner (1976), for example, proposed that a sequence in southern Sweden might be the standard reference point for the Pleistocene–Holocene boundary. The third view, and the one which has been most widely adopted, is that the Holocene is simply defined as beginning 10 000 ^{14}C years ago.

However, there is a difficulty in knowing what 10 000 ^{14}C yr BP actually means in terms of real calendar ages. ^{14}C-dated tree-ring sequences have shown a systematic divergence between the two timescales which amounts to 1000 years or more by the beginning of the Holocene. Moreover, ^{14}C time is elastic; that is, it has been stretched or compressed at different periods in the past. And perversely – if not coincidentally – 10 000 ^{14}C yr BP happens to coincide with a major ^{14}C 'plateau', when the ^{14}C 'clock' stood still, while at least 400 years of real time elapsed (Taylor et al., 1996) (figure 2.3). This ^{14}C plateau was probably caused by abrupt changes in atmospheric CO_2 and oceanic circulation at the end of the last glaciation (Goslar et al., 1995), and it has meant that 10 000 ^{14}C yr BP represents not one moment in time, but several! It would clearly be preferable to define the onset of the Holocene in calendar and not in ^{14}C years. But which date should be chosen? There is evidence of rapid climatic change around this time in a number of independently dated natural archives, and the different calculated ages for the change are shown in table 2.1. These different techniques do not provide identical results, but most of them lie between 11 000 and 11 800 Cal. yr BP. The CALIB 3.0 program gives c.11 200 Cal. yr BP as the calibrated equivalent of 10 000 ^{14}C yr BP, although this is a little younger than counts of annual snow layers in Greenland ice cores (Alley et al., 1993). Following Gulliksen et al.

(1998), an age of 11 500 Cal. yr BP will be used for the start
of the Holocene, although it is recognized that this age could
be in error by up to 300 years either way.

Table 2.1 Calendar age estimates for the beginning of the
Holocene

Age (Cal. yr BP)	Method
11 200	Tree-ring calibration of ^{14}C on German pines
11 360–11 920	Counting of annual layers in GISP2 Greenland ice core
11 440–11 580	Counting of annual layers in GRIP Greenland ice core
11 200–11 700	U-Th calibration of ^{14}C in tropical corals
11 395–11 540	'Wiggle matched' ^{14}C lake sediment record, Kråkenes Lake, Norway
11 360–11 600	Varve calibration of ^{14}C, Lake Gosciaz, Poland
c.11 490	Varve calibration of ^{14}C, Holzmaar, Germany
10 960–11 100	Varve calibration of ^{14}C, Soppensee, Switzerland
11 440	Varve counting, Sweden

Sources: Alley et al. (1993), Björck et al. (1996), Gulliksen et al.
(1998)

problem is that ^{14}C calibration beyond about 11 000 Cal. yr BP
is far from universally agreed. Different dating techniques have
produced significantly different age estimates for the start of
the **Late-glacial** period, for instance. None the less, all agree
that ^{14}C and calendar ages continue to diverge back to the last
glacial maximum. For our purposes, they cross-match sufficiently well to make it possible to use a calibrated timescale
for all of the events and changes covered in this book, even if
the precision achieved for ^{14}C calibration is much reduced
before about 10 000 years ago.

Other incremental dating methods

Dendrochronology is only one of a number of incremental
dating methods. One potentially precise technique involves
counting the annual layers of snow that build up to form ice
sheets. If the accumulation significantly exceeds the summer
melting, and if it is not deformed by lateral movement (as it is
in a valley glacier), then the snow layers will be preserved
even after later compaction turns them to ice (see plate 3.1,
p. 64). After a certain point, these layers become so squashed
by the overlying weight of ice that they are no longer visible.

None the less, in the GISP2 Greenland ice-sheet core it was possible to count every annual layer back to 17 400 Cal. yr BP with an estimated precision of ±3 per cent (Alley et al., 1993).

In similar fashion, lake sediments may be finely layered, or laminated, representing one year's accumulation. Annually laminated lake sediments are known as **varves**, and can form in a number of different ways. The varves studied by the Swedish scientist Baron de Geer at the end of the nineteenth century had formed in lakes adjacent to the former Scandinavian ice sheet, and comprised couplets of alternating coarse and fine-grained layers (Sturm, 1979). These clastic varves provided one of the first quantitative estimates for the duration of the Holocene. On the other hand, laminated lake sediments can form wherever there is a strong seasonal variation in sediment supply or where circulation of lake water is absent or incomplete for at least part of the year (O'Sullivan, 1983). One of the commonest situations in which non-glacial varves form is in lakes that freeze over in winter, as is the case in much of North America and Scandinavia. Organic matter accumulates in the still-water conditions beneath the ice cover, often followed in spring by a deposition of **allochthonous** sediment washed into the lake during snow-melt. In the summer months photosynthesis by algae may lead to a layer of **diatoms** settling out on the lake bed. Varves can also be chemical in origin if the lake water is either strongly acid or alkaline, or if it becomes saline. Alkaline lakes such as Zürich See in Switzerland are rich in carbonate, which summer photosynthesis causes to be precipitated out as a $CaCO_3$ white-out (Kelts and Hsü, 1978). Varves have proved useful even when they are not continuous to the present day and are therefore free-floating in time, for example, in calculating the length of the cold Younger Dryas **stadial** at the end of the last ice age (Lotter, 1991).

Using LICHENS for dating achieves a very different degree of precision. Lichens represent a symbiotic relationship between fungi and algae. The algae photosynthesize to provide nutrition for the partnership and the resulting compound organism behaves as if it were a single biological unit. Lichens are able to exploit a wide range of habitats and grow on substrates ranging from tree trunks to bare rocks. Because of their ability to survive in unfavourable environments they often take the role of primary colonizers in plant succession. Their utilization for dating purposes exploits the fact that most lichen species grow at rates that are both fixed and measurable. The larger the lichen, the greater its age. Empirical work by R.E. Beschel in arctic and alpine environments during the 1950s and 1960s showed there to be a clear relationship between lichen size – as measured by diameter – and age. Beschel went on to formalize this as a dating technique which has subsequently been applied to many field studies, notably of Holocene glacier fluctuations (Innes, 1985). Lichenometry is especially valuable in glacial studies because of the difficulty of applying other dating techniques to such problems as establishing a date for exposure of a bare rock surface following deglaciation (Matthews, 1992).

Field techniques for the measurement and sampling of lichens for dating are relatively straightforward (Lock et al., 1979), but doubts have been raised about whether the assumptions behind them have been adequately tested (McCarroll, 1994). In particular, it remains uncertain how far lichen growth rates are affected by variables other than time, including competition for space, light and nutrients. In consequence, lichenometry is best considered as a dating method that can provide relative rather than absolute ages.

Palaeomagnetic dating

The Earth acts as a giant magnet with its own magnetic field, and this field changes constantly. Perhaps the most dramatic example of these changes is the periodic reversals in the Earth's magnetic polarity. However, the last time the compass needle slipped from the south to the north pole was over 700 000 years ago, and although claims have been made for more recent short-lived magnetic excursions, polarity reversals occur too infrequently to help date the Holocene. On the other hand, so-called secular magnetic variations can be important in helping to provide a timescale over hundreds or thousands of years. It has long been recognized that magnetic north does not coincide with geographic or true north, and that the position of the former has wandered through time. Measurements of the geomagnetic field have been taken at London for over 400 years and show that magnetic north migrated westwards between AD 1580 and 1820, when it lay 20°W of true north. Since then it has shifted eastwards once more so that today it lies at about 11°W.

The angle between true and magnetic north, or declination, is only one of three components of the magnetic field. The compass needle will not only move from side to side but, if freely suspended, dip down at an angle towards the Earth. This dip, or inclination, varies from the equator to the poles, as does strength or intensity of the Earth's magnetic field. All three of these components vary through time and in combination produce a characteristic palaeomagnetic signature. Their magnetic alignments can be incorporated and preserved in baked materials or in sediment particles which settle out in standing water. The study of archaeomagnetic samples such as hearths has extended the record of secular magnetic variations

Figure 2.4
Palaeomagnetic dating
curves for South
Australia (data from
Barton and
McEllhinny, 1982)

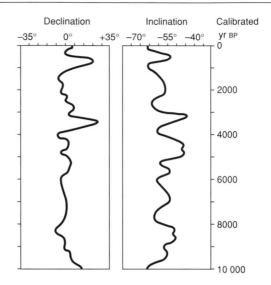

back to before AD 1000 (Aitken, 1990). For earlier periods magnetic measurements have been applied to wet lake sediment cores with considerable success (Thompson and Oldfield, 1986). Secular magnetic variations do not, on their own, provide age estimates. However, once a master geomagnetic curve has been dated by another method such as ^{14}C, it can be applied to sites elsewhere. Holocene master curves, which are regional rather than global in extent, now exist for six areas of the world (Thompson, 1984). Those for South Australia come from the crater lakes Bullenmeri and Keilambete and are shown in figure 2.4. Measurement by magnetometer is rapid and, in the case of declination, non-destructive. On the other hand, not all lakes are equally well suited for the technique, which is effectively restricted to lake sediments with homogeneous particle size and organic content down core.

Conclusion

Possibly the most notable feature of dating techniques is the great diversity of ways in which age estimates can be obtained for the Holocene. Diversity is the key in other respects as well; dating techniques have different age ranges over which they can operate; they cannot all be applied to the same sample materials; some destroy the sample during dating, while others do not; and so forth (see table 2.2). Overall dating reliability is consequently improved by the use of several techniques in combination. Some of the most precise dates, other than historical ones, come from tree-rings. These are accurate as well as precise, but can be relatively time consuming to obtain. Lake varves and ice layers can achieve a similar precision of one

Table 2.2 Dating methods applicable to the Holocene

Dating method	Time range Cal. yr BP	Precision yr	Applications
Historical			
archival	1–5000	1–50	varied; e.g. glacier histories
archaeological	50–>40 000	1–1000	varied; e.g. alluvial histories
Radiometric			
radiocarbon	200–40 000	20–1000	organic materials
luminescence	50–>40 000	50–5000	burnt clay and pottery, most minerogenic sediments (e.g. loess)
uranium series	1000–>40 000	20–5000	carbonates (e.g. coral), volcanics
lead-210	10–200	1–10	lake and coastal sediments, peat
caesium-137	1–35	1–5	lake sediments
Incremental			
tree-ring	1–11 000	1–5	wood timber
varves	1–20 000	1–10	lake and some marine basin sediments
annual snow/ice layers	1–>20 000	1–350	ice-sheet cores
lichenometry	10–5000	10–1000	glacial histories
Palaeomagnetism			
secular variations	100–15 000	10–100	lake sediments

year or less, and have the additional advantage of dating deposits which are prime sources of palaeoenvironmental information. On the other hand, none of these is as widely applicable as ^{14}C dating.

Dating is not only a matter of precision and accuracy, but also of attribution and interpretation. Dated samples almost always come from a stratigraphic context of one kind or another, and after age determination they must be related back to the site stratigraphy. Because ^{14}C and ^{210}Pb, in particular, provide spot dates rather than a continuous chronology, the age of intervening stratigraphic layers needs to be assessed by interpolation between or extrapolation beyond dates obtained by these methods. This is normally achieved by use of an age-depth curve, in which dates are related to their stratigraphic position in a sediment core or profile (see figure 2.5). In sediments which are still accumulating, the uppermost layers can be assumed to date to the present; otherwise at least two dates are necessary for an age-depth curve to be drawn up.

If change through time at a single sequence is investigated via stratigraphy, the matching of different sequences over space is achieved via **correlation**. Features such as the European elm decline are best demonstrated to be of the same age if they are dated independently at every site. Correlation using pollen or other biostratigraphic evidence is likely to be imprecise and

Figure 2.5 Age-depth curve and accumulation rates for a lake sediment core with three ^{14}C dates

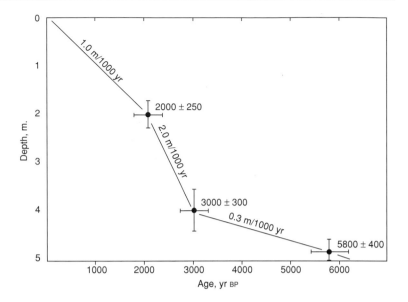

may lead to circular reasoning, but other means of correlation can be a reliable alternative to independent dating. Prominent among these is the use of volcanic ash layers, or **tephro-chronology**, as marker horizons. Fossil soils and landforms can also sometimes be traced over considerable distances and are normally age-equivalent rather than time-transgressive.

Palaeoecological Techniques

Palaeoecology is the study of fossil organisms in order to re-construct past environments. Accurate palaeoenvironmental reconstruction also depends on modern ecological data, and is consequently only as good as our knowledge of present-day ecosystems. A fundamental assumption made here is that rela-tionships which can be observed at the present time also held good in the past. This principle is known as **uniformitarianism**. Take the case of *Hippopotamus* bones found in the middle of the arid Saharan desert and dated to between 10 000 and 6000 years ago. Assuming that the bones have not been moved sig-nificantly since death, then two possible explanations might be offered. The first and more obvious would be to employ our knowledge of modern hippo ecology to infer that the Sahara was a land of lakes and rivers during the early Holocene. The alternative explanation would be that hippos have changed their habits and habitats, formerly being happy scaling sand dunes and only recently taking to wallowing in mud.

In this example the case for the present not being the key to the past seems somewhat absurd. But for organisms with

1 We understand the environmental factors governing present-day plant and animal distributions.	**Table 2.3** Uniformitarian* assumptions in palaeoecology
2 Their ecological affinities have not changed through time.	
3 Present (and past) distributions are (and were) in equilibrium.	
4 Modern analogues exist.	
5 The origin (taphonomy) of a fossil assemblage can be established, and that . . .	
6 . . . this assemblage is not biased by contamination or differential preservation.	
7 Fossils can be identified to a meaningful level of taxonomic resolution.	

* *sensu* actualism

short life cycles, such as beetles (Coleoptera), the validity of the principle of uniformitarianism may be called into question. Up to 20 000 generations of beetles have lived and died since the last glacial maximum, perhaps sufficient for evolutionary changes to have taken place. In fact, detailed comparisons of modern and ice-age beetle skeletons have revealed no discernible changes in their anatomy, indicating evolutionary stability (Coope, 1970). Even so, uniformitarianism remains assumed not proven, and the further back in time we go, the more uncertain this assumption becomes. Uniformitarianism remains an important issue in other ways too; for example, in the problem of interpreting an assemblage of fossils which has no modern equivalent, or analogue (see table 2.3).

The process by which a group of living plants or animals is transformed into an assemblage of fossils is by no means always a straightforward one. This transformation and the subsequent interpretation of the fossil assemblage is illustrated here with reference to pollen analysis. Pollen analysis, or palynology, is only one of many branches of palaeoecology, but it is broadly representative of the wider subject in terms of the methods it employs.

Principles of pollen analysis

Palynology is the single most important branch of palaeoecology for the late Pleistocene and Holocene. It has been extremely widely applied since it was first applied in stratigraphic studies by the Swede Lennart von Post at the start of the twentieth century (Faegri and Iversen, 1989). Its basis is that pollen grains and spores produced as part of plant reproduction may be incorporated and preserved in lake muds, peat bogs or other sediments which can later be analysed to reconstruct the vegetation of an area.

Table 2.4 Relative pollen productivity of selected temperate trees	× 1	lime (*Tilia*), ash (*Fraxinus excelsior*), maple (*Acer*)
	× 2	beech (*Fagus*)
	× 4	elm (*Ulmus*), spruce (*Picea*)
	× 6	hornbeam (*Carpinus*)
	× 8	alder (*Alnus*), birch (*Betula*), hazel (*Corylus*), pine (*Pinus*), oak (*Quercus*)

Source: Andersen (1973)

Pollen production starts with a life assemblage, such as mixed temperate woodland. Within the woodland ecosystem, pollen is produced more abundantly by some plants than by others (see table 2.4); in particular, productivity is much greater for wind-pollinated than for insect-pollinated taxa. Of the many pollen grains produced, only a few are used for reproductive purposes, with the remainder being dispersed through the environment. Pollen dispersal involves several pathways, notably via high- and low-level air-borne transport and via a water-borne route in runoff and streams (Tauber, 1965). Each pathway introduces its own selective bias into the dispersal process. High-level wind dispersal, for example, leads to a well-mixed pollen 'rain' with the pollen of some taxa being carried over long distances, notably the conifers (gymnosperms), whose pollen grains have sacs to help keep them air-borne.

The processes of pollen production, dispersal and deposition thus generate a death assemblage which is significantly different in its composition from the initial life assemblage. This is further altered by sediment diagenesis and by differential fossil preservation and destruction (Birks and Birks, 1980, ch. 1). Pollen is best preserved under **anaerobic**, typically acid conditions, such as are encountered in blanket peat bogs. The by-now fossil assemblage can be investigated through field and laboratory work, typically involving coring at a field site (see plate 2.3) followed by laboratory preparation of samples. In order that only the sporopollenin remains intact, inorganic components of the sediment such as carbonate and silica are removed by reagents like hydrochloric and hydrofluoric acids. After mounting on a glass slide, pollen grains can be examined under the microscope, usually at magnifications between ×400 and ×1000. Individual pollen taxa have different shapes, sizes, number of apertures and other features which allow them to be differentiated (Moore et al., 1991). Identification is normally to the level of genus for trees and shrubs, and only to family for many herbs and grasses: species-level identification is the exception rather than the rule in palynology.

We can illustrate the sequence described above by means of the example shown in figure 2.6. In this hypothetical case only four plant taxa are present; pine (*Pinus*) and deciduous

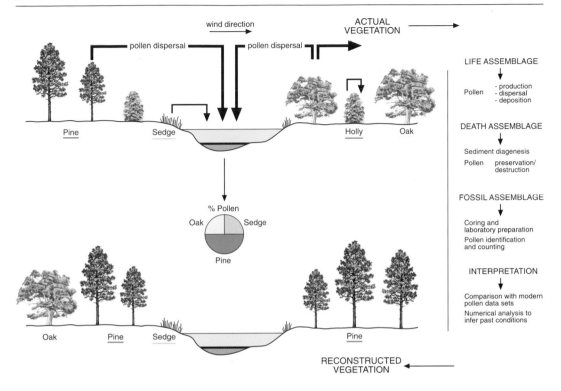

Figure 2.6 Vegetation reconstruction based on pollen analysis (see text for explanation)

oak (*Quercus*) form separate woodland stands either side of a lake, holly (*Ilex aquifolium*) is an understorey shrub, and there is a fringing mat of sedges (Cyperaceae) around the lake. Although pine and oak produce abundant pollen, the prevailing wind direction causes pine to be more strongly represented in the pollen rain reaching the lake. In contrast, holly is hardly represented at all, because it is insect-pollinated. The pattern of vegetation reconstructed from fossil pollen includes aspects of the original flora, but these have been modified and are incomplete. Not only does pine appear more abundant than oak, but it is difficult to know if the two types of tree formed separate or mixed woodland stands. The sedge pollen could also be interpreted in more than one way, for while some species form sedge swamps, other species occupy very different habitats such as tundra. The rarity of holly pollen grains might cause it to be overlooked altogether in the reconstruction. It is precisely in order to overcome some of the biases illustrated here that palaeoecologists carry out studies of modern pollen deposition and preservation as an integral part of their work.

The reconstruction shown in figure 2.6 is based on the relative proportions of different pollen taxa. To obtain representative proportions it is normally necessary to count at least 200 pollen grains in any sample. However, even if a much larger

Table 2.5 Absolute and relative pollen-counting statistics

	Absolute numbers			Relative proportions		
	pine	*oak*	*sedge*	*pine*	*oak*	*sedge*
time 1	10	5	5	50%	25%	25%
time 2	10	5	35	20%	10%	70%

number than this were to be counted, the 'relative' approach has certain inherent limitations. Imagine that the lake in our example was progressively infilled with sediment, and that as it became shallower so sedge swamp largely replaced open water. Tree pollen would be outnumbered by a super-abundance of locally produced sedge pollen (see table 2.5). In a pollen count based on relative proportions, pine and oak would appear to decline in importance between times 1 and 2 even though the surrounding woodland vegetation was in fact unchanged. In an attempt to overcome this difficulty palyno-logists often exclude some local, wetland pollen types from the total pollen sum on which percentages are based. On the other hand, it may be difficult to assign particular pollen types to a local wetland as opposed to a dryland origin with certainty. An alternative approach is to count 'absolute' numbers of micro-fossils to obtain pollen concentration per cm^3 of sediment. This is most often done by adding a known quantity of exotic markers, such as *Lycopodium* spores or polystyrene micro-spheres, into the sample during preparation. If a timescale is available for the site under study, concentration values can be translated into rates of pollen influx per year.

Most pollen analyses are carried out at sites – such as lakes – which have experienced more or less continuous sediment accumulation. A series of pollen samples from different depths in the sediment profile can then be presented as a pollen dia-gram, which conventionally shows depth or age on the verti-cal axis and frequency on the horizontal axes. Pollen and other microfossil diagrams are used extensively in this book, in some cases based on absolute counts, in others based on propor-tional ones. The diagrams sometimes show only selected taxa, or major pollen groups such as arboreal (tree) versus non-arboreal pollen (AP *vs* NAP). Figure 2.8A is a percentage pollen diagram from a site in southern France, which shows changes in selected important tree types against core depth. As the calibrated ^{14}C dates indicate, this section of the core covers the transitional period from the Late-glacial stage through to the middle of the Holocene. Like most pollen diagrams, it is subdivided into zones within which samples (or spectra) are similar in their essential characteristics. Before the advent of radiocarbon dating, these zones were assumed to be synchronous over whole regions. In the British Isles, for instance, Sir Harry Godwin proposed a pollen zonation scheme that began in the Late-glacial (zones I–III) and continued through the Holocene to the present day (IV–VIII). Regional pollen zones are now known to be time-transgressive and are therefore used less widely than previously, but local pollen assemblage zones remain an essential part of individual site investigations. Increasingly, these local pollen zones are designated with the help of statistical methods such as cluster analysis (Birks, 1986b).

Pollen, climate and human impact

The initial objective of a pollen analytical study is normally to reconstruct past vegetation, but there is often a second objective – that of establishing the factors which determined the former flora. The most important of these controlling factors are CLIMATE and HUMAN ACTIVITIES. Assuming little human disturbance, it is widely believed that vegetation will eventually achieve a state of equilibrium with the prevailing climate (Webb, 1986; Edwards and MacDonald, 1991a; Ritchie, 1995). Modern data on climate–plant relationships may therefore be applied to pollen-based vegetation histories to produce estimates of past temperatures or rainfall levels (Birks, 1981). In West Africa, for example, the montane forest tree *Olea hochstetteri* is today restricted to land above 1100 m. Its pollen is prominent in the bottom part of a diagram from the lowland site of Lake Bosumtwi, and this has been used to infer that prior to 8500 years ago temperatures here were at least 2–3°C colder than at present (Maley and Livingstone, 1983).

Comparable northern European examples are the arctic/alpine species, dwarf birch (*Betula nana*) and mountain avens (*Dryas octopetala*), whose pollen or macrofossils are common in Late-glacial assemblages. In the British Isles, dwarf birch is today virtually restricted to the Scottish Highlands, where mean summer temperatures are below 22°C (Connolly and Dahl, 1970). But before accepting this as the temperature of a typical July day in southern England 11 000 years ago, it should be remembered that species distribution is a function of more than one climatic variable. In an attempt to overcome this problem, Johannes Iversen (1944) combined two climatic variables and several different indicator species. He showed, for example, that holly can tolerate cool summers but not cold winters, whereas mistletoe (*Viscum album*) is the reverse – requiring summer warmth but being relatively hardy in winter. Ivy (*Hedera helix*) lies somewhere between the other two

Plate 2.3 Coring at the edge of lake Gölhisar in southern Turkey. The resulting 8.3 m-long core produced a pollen record spanning the whole of the Holocene

in its climatic tolerances (see figure 2.7). Using the overlapping ranges of the three shrubs in combination it is possible to provide more precise temperature reconstructions than if each were used on its own, and to provide an indication of past climatic continentality.

To extend this approach to a 'multivariate' analysis of vegetation and climate requires the use of more complex numerical methods, possible only with the help of a computer. Tom Webb and colleagues (Webb, 1985; Bartlein et al., 1986) developed numerical, multivariate techniques for reconstructing

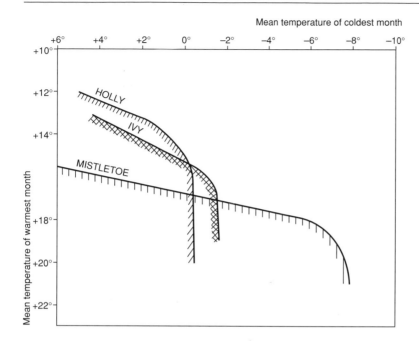

Figure 2.7 The thermal limits of holly, ivy and mistletoe (after Iversen, 1944)

climate directly from pollen data. This collapses the normal two-step reconstruction of pollen to vegetation to climate into a one-step process, involving the creation of so-called pollen–climate 'response surfaces'. Climatic calibration of modern and fossil pollen data has allowed maps to be produced showing temperature, precipitation or air mass distribution for different times in the past over eastern North America (Webb et al., 1993). This procedure cannot be applied so easily in Europe because human impact on vegetation has a much longer antiquity here, and mapping work has concentrated more on pollen or vegetation than on climate (Huntley and Birks, 1983; Prentice et al., 1996).

The interpretation of pollen records in terms of past human impact is, if anything, more difficult than it is for climate (Edwards and MacDonald, 1991b). Many former land uses, such as hunting-fishing-foraging in Mesolithic Europe, have no modern analogue for comparison. Furthermore, it is often the indirect consequences of human impact that are most obvious palynologically (Edwards, 1979, 1982). The pollen of wheat, barley and other cereal crops is poorly dispersed and only weakly represented in European pollen diagrams until historical times. More prominent evidence of prehistoric agriculture instead comes from the pollen of ruderals (weedy plants) like docks (*Rumex* sp.) and nettles (*Urtica* sp.) (Turner, 1964; Behre, 1981, 1986). These species take advantage of disturbed ground regardless of cause, and are therefore not diagnostic indicators of anthropogenic disturbance. Many, for

example, were relatively abundant during the Late-glacial before a stable forest cover had been established (Godwin, 1975;
Huntley and Birks, 1983, p. 518ff.).

The recognition of human disturbance in pollen diagrams
usually involves a combination of evidence, including changes
in the overall composition of vegetation as well as indicator
species (Maguire, 1983; Birks et al., 1988). Amongst the latter
would be grassland indicators such as ribwort plantain
(*Plantago lanceolata*), whilst the former would include changes
in the AP–NAP ratio. Where several lines of evidence converge
– as in the pollen diagram shown in figure 6.11 (see p. 190) –
interpretation is likely to be straightforward. In other cases a
former vegetation change could be attributable to more than one
cause: the mid-Holocene European elm decline, for example,
has been variously attributed to prehistoric agriculture, disease,
climate change, or a combination of more than one of these
factors. Equifinality is the term used when a similar end-result
could have been produced by more than one different cause. It
is one of the tasks of palaeoecology to test between competing
causal hypotheses, for instance in explaining what has caused
many Scandinavian lakes to be acidified (see chapter 7).

Plant macrofossils and charcoal

Notwithstanding the enormous success of palynology in reconstructing former vegetation, climate and land use, pollen
research alone cannot hope to answer all our questions about
past ecologies. In some situations other sources of proxy data
may prove more informative, and the potential range of biological indicators is very wide – from microscopic algae to
mammoths. On calcareous rocks where pollen sites are almost
absent, such as English chalk, land snails (Mollusca) have
proved of great value in the investigation of land use and
vegetation history (see Technical Box VII, pp. 196–7). Beetles
(Coleoptera) are another important type of organism which
can respond rapidly and sensitively to changes in climate (see
Technical Box III, pp. 69–70). In **palaeolimnological** studies
aquatic indicators such as diatoms (see Technical Box IX,
pp. 230–1) are especially valuable. The remains of many types
of micro-organisms can furnish valuable palaeoecological information in different contexts, ranging from Protozoa such
as Foraminifera (calcareous marine zooplankton) and testate
amoebae (Warner, 1988), through ostracods (Holmes, 1992) to
the pigments left by phytoplankton (Leavitt, 1993). The value
of alternative, complementary sources of information is well
illustrated by the example of plant macrofossils, that is, plant
remains like seeds, leaves and wood that are visible with the
naked eye.

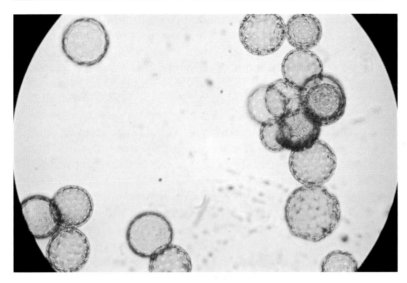

Plate 2.4 Pollen grains of *Chenopodium album*, an indicator of open-ground conditions, *c*.40 microns (0.04 mm) in diameter

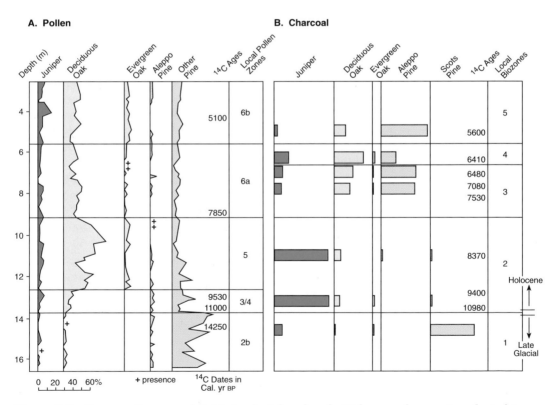

Figure 2.8 Comparative records of Late-glacial and early Holocene changes in selected trees from southern France; **A** pollen record from a valley peat bog at Tourves (after Nicol-Pichard, 1987); **B** wood charcoal record from prehistoric rockshelter site of Fontbrégoua (data from Thiébault, 1997)

Most of the same principles examined in relation to pollen – life and death assemblages, and so forth – apply also to plant macrofossils, and indeed to other organisms. However, bigger sample volumes are required because macrofossils are larger than pollen, and counting statistics are more difficult because of the wider variety of parts preserved. (This applies even more to bone remains of mammal, reptile and bird faunas.) Compared with pollen, however, plant macrofossils have a number of advantages (Birks and Birks, 1980, p. 66ff.). They form part of the once-living plant rather than the reproductive stage of the life cycle. As a result, they are usually identifiable to species level and the ecological inferences that can be drawn from them are relatively precise. Unlike pollen grains, macrofossils are not transported far from their point of origin and their local presence can be assumed with greater confidence. They can also help record plants with poor pollen production or dispersal, such as the arctic flower, mountain avens (*Dryas octopetala*).

On archaeological sites macrofossils are often found in the form of charred seeds or wood charcoal, which offers a direct indication of the plant resources exploited by former human communities (Helbaek, 1969; Vernet and Thiébault, 1987). The study of wood charcoals – or anthracology – can offer useful insights into localized vegetation changes which complement those provided by pollen analysis (Chabal, 1992). Figure 2.8 compares the Late-glacial to mid-Holocene woodland history from a charcoal record in an excavated prehistoric rockshelter in southern France with a pollen core from a valley mire nearby. The latter represents the more continuous record of regional vegetation change, but there are some problems with interpretation of pollen data which the charcoal can help to resolve. For instance, the Late-glacial period before *c.*11 500 Cal. yr BP has high pine pollen percentages at Tourves, but does this really mean that pine trees were common in the vicinity? Pine is notoriously well dispersed by the wind and often makes up a significant proportion of the total pollen found in open, exposed landscapes such as tundra, even though it may not be growing locally (Ritchie, 1987). Fortunately, we know in this case that pine trees really were present – and Scots pine at that – because the Epi-Palaeolithic occupants of the rockshelter collected it at the time (figure 2.8B). On the other hand, changes in the charcoal record can also have been caused by cultural preferences, say, because one type of wood burns better than another. The switch from juniper to oak and pine charcoal just before 7500 Cal. yr BP, for example, coincides with the appearance of the first (Neolithic) farmers in this area. This apparent vegetation change may reflect the fact that oak and pine woods were being cut down to make way for cultivated crops, rather than because juniper – which had been used previously – became less common in the regional vegetation. Charcoal particles are also found in lake sediments and peat bogs (Patterson et al., 1987), but as microscopic, rather than macroscopic, fragments. While they are normally too small to permit identification to tree type, these tiny charcoals do offer very important information about fire histories (e.g. Bennett et al., 1990).

Plant macrofossil analysis has been applied to a range of palaeoecological problems. In western Norway, as in Provence, there were some striking differences between the vegetation records indicated by pollen and plant macrofossils, in this case from lake sediment cores (Birks, 1993). In particular, birch (*Betula*) was prominent in the pollen diagrams, but absent in the macrofossil record from the same lakes during a Late-glacial stadial. As the latter normally provides the more reliable index of local conditions, we can assume that birch was not, in fact, growing in the vicinity of the lake, and that its pollen must have been blown in from further afield. One task for which plant macrofossils are well suited is reconstructing the history of wetlands as they progress through a **hydrosere** succession from open water through sedge swamp and fen carr to peat bog (Walker, 1970). Because peat deposits are almost entirely made up of plant macrofossils, studies of peat stratigraphy and of macrofossils often amount to the same thing. The relative importance of the remains of *Sphagnum* moss and heathland plants (e.g. *Calluna*, *Eriophorum*) in peat bog sequences provides an indicator of surface wetness and hence changes in hydro-climatic conditions (e.g. Barber, 1981).

A very different application of plant macrofossil analysis has been developed in the arid lands of the American Southwest. Here, the desert packrat (*Neotoma* spp.) collects plant material from within a 30 m radius to form midden deposits (Wells, 1976; Betancourt et al., 1990) rather as the beaver (*Castor*) does in the boreal forest. These middens represent a sample of the local vegetation, which is subsequently preserved by being soaked and cemented by packrat urine! The packrat's somewhat unhygienic habits have given us an excellent record of vegetational and climatic history. The plant macrofossils that make up the packrat middens are not only identifiable to the level of species, but can also be dated directly by radiocarbon. They have been used, for instance, to show that during the late Pleistocene the Sonoran and southern Mohave deserts supported woodland of pinyon pine (*Pinus monophylla*) and other pygmy conifers, not the creosote-bush scrub existing today (Spaulding et al., 1983). This implies at least 400 m lowering in the altitudinal range of coniferous trees prior to

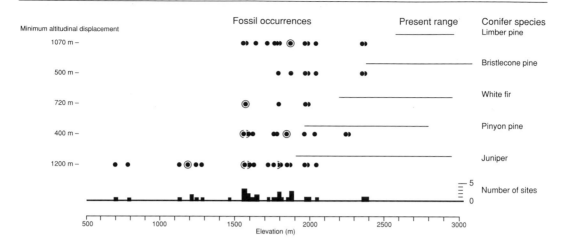

Figure 2.9 Present-day and late Pleistocene altitudinal ranges of coniferous trees in the Mohave desert based on packrat middens (after Spaulding et al., 1983). Circled dots indicate an occurrence at more than one site at that elevation

13 000 Cal. yr BP (see figure 2.9). Middens formed by hyrax (Procaviidae) and similar 'desert rats' are found in Africa, southwest Asia and Australia, opening the possibility of applying more widely this remarkable source of palaeoecological data (Scott; 1990; Fall, 1990; Nelson et al., 1990).

Geological Techniques

Recent Earth history can be understood by examining sediment sequences and landforms at the Earth's surface. These can be classified into a series of different geomorphological regimes, each with its own characteristic set of depositional environments (see table 2.6). As in palaeoecology, the reconstruction of past geological environments is based on uniformitarian principles; in other words, our knowledge of the present is used as the key to understanding the past. Modern data about Earth surface processes are applied to fossil landforms and sediments, first, to assign them to a particular geomorphological regime,

Table 2.6 Characteristics of different geomorphic regimes

Geomorphic regime	Sedimentation	Sediments	Landforms
Lacustrine	continuous (lake bed)	lake marl	shoreline terrace
Fluvial	semi-continuous	river gravel	palaeo-channel
Coastal marine	semi-continuous	beach sand	raised beach
Aeolian	semi-continuous	dune sand	fossil dune
Glacial	discontinuous	till	end-moraine
Mass movement	discontinuous	'head'	solifluction lobe

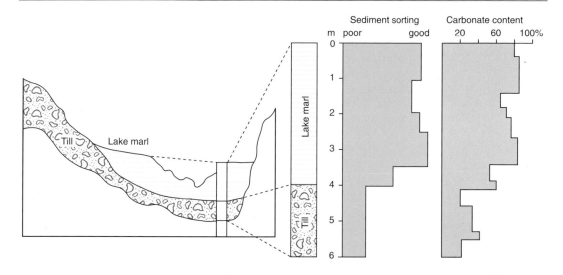

Figure 2.10
Environmental
reconstruction based
on landform and
sedimentary evidence

and, second, to interpret their origin in terms of past processes and climates. No-analogue situations are encountered less often than in palaeoecology, although they do sometimes occur, notably for high-magnitude events. The so-called 'Channelled Scabland' of Washington, for instance, was for many years explained in terms of observable modern fluvial processes (Baker and Bunker, 1985). It is now realized that it was in fact created by catastrophic, short-lived flooding when ice-dammed Lake Missoula burst at the end of the last glaciation.

LANDFORMS may be produced by erosion, deposition, or a combination of both types of process. Terrace features, in particular, are formed by a phase of sedimentation – say, by an aggrading river – followed by a period of erosional downcutting. Terraces provide valuable information on previous river, lake and sea levels that can be recorded by survey and mapping. However, it is possible to misinterpret the origin of terraces and other landforms unless data on surface morphology are supplemented by an analysis of the underlying sediments. Earth histories can therefore be studied most effectively by combining landform-based techniques with those based on sediment stratigraphies (Goudie, 1994), as illustrated by the sequence shown in figure 2.10. Here a glacial moraine is overlain by lake sediments, now being dissected by fluvial erosion. Of these three geomorphic regimes, only the first two have left sedimentary records, and their contrasting physical and chemical characteristics are recorded on one side of the diagram.

Minerogenic or clastic sediments, such as river gravels, comprise rock particles of allochthonous origin – that is, they were brought to the point of deposition from elsewhere. Analysis of their physical characteristics involves examining particle shape, size distribution and arrangement in space. One standard

Table 2.7
Comparative
properties of glacial
and aeolian sediments

	Glacial till	Dune sand
particle shape	angular	spherical
grain surface	fractured	smooth
size distribution	mix of fine and coarse particles	homogenous sand
particle sorting	very poor	good
sediment structures	weak or absent	obvious cross-bedding
particle alignment	may be parallel to direction of ice flow	not evident

measurement technique is particle size analysis, carried out by sieving, pipette analysis and other methods, from which size and sorting indices may be derived (Folk, 1974). The characteristics of two contrasting clastic sediment types are listed in table 2.7. Till and dune sand are easily distinguished from each other, but differentiation may be harder in other cases. Soliflucted 'head' has similar properties to some tills, for instance, while desert dune and beach sand could be confused on the basis of sediment characteristics alone. Minerogenic deposits can also be analysed for their clay and 'heavy' mineral compositions, and for their mineral magnetic properties.

PRECIPITATES, such as lake marl, are sediments chemically deposited from aqueous solution and are therefore **autochthonous** in origin. Most commonly, deposition takes place in standing-water conditions such as are encountered in lakes and the sea, and involves one of two main mechanisms. The first of these is biochemical 'fixing' of solutes by aquatic or marine organisms, like the polyps responsible for building up coral reefs. The second is direct chemical precipitation from salinealkaline water where solutes are highly concentrated, for example in a playa lake or coastal lagoon (see plate 2.5). Different types of salt are deposited from different solutions, depending on the chemical composition and concentration of the water (Eugster and Kelts, 1983). Another type of chemical precipitate is speleothem carbonates found in caves, including stalagmites and flowstones. Like other precipitates, speleothems can be analysed isotopically (see below).

Biogenic sediments comprise primarily organic carbon and are most often autochthonous, for example in a raised peat bog. More rarely, organic matter may be allochthonous, as when soils are eroded and washed into streams and lakes. Biogenic sediments typically contain abundant micro- and macrofossils and they are therefore extensively used in pollen and other palaeoecological investigations.

In practice, most sediments comprise a mixture of clastic, chemical and biogenic elements. One way in which the differ-

(a) *Substantia humosa*: undifferentiated disintegrated organic matter

(b) *Turfa*: macroscopic plant remains of below-ground origin and mosses; e.g. roots

(c) *Detritus*: macroscopic plant remains of above-ground origin; e.g. leaves

(d) *Limus*: lake mud of biogenic or chemical origin; e.g. marl

(e) *Argilla*: fine-grained minerogenic particles; e.g. clay

(f) *Grana*: medium or coarse-grained minerogenic particles; e.g. sand

Table 2.8 Elements of the Troels-Smith sediment classification

ing proportions of elements making up a sediment sample can be assessed is by using a system of sediment classification, such as that developed by the Danish palaeoecologist J. Troels-Smith (Birks and Birks, 1980, p. 38ff.; Aaby and Berglund, 1986). Of Troels-Smith's six main sediment categories, three are biogenic, a further two are minerogenic, and a further one is either chemical or biogenic (see table 2.8). Although this system of sediment classification is widely applied, its usage is restricted to freshwater lake and mire sediments in humid temperate regions. Lakes and mires certainly provide some of the most accessible and best-studied sediment 'archives' recording Holocene environmental change, with the former being especially informative when the sediments are annually laminated, or varved. Sediment in small temperate-zone lakes and mires typically builds up at about 1 m per thousand years, so that a full Holocene record is likely to be represented by around 10 m of accumulated sediment (Webb and Webb, 1988).

Lakes and mires are, however, but two among many types of stratigraphic record. The deep ocean bed, for example, is a depositional environment which has proved invaluable for understanding the Earth's history during the Pleistocene ice age. Deep-sea sediment cores contain a record of long-term fluctuations between cold glacial and warmer interglacial climates (Lowe and Walker, 1997). The slow rate of sedimentation is ideal for studying such long spans of time, although it means that the uppermost sediments formed during the Holocene are typically no more than 50 cm thick. This does not allow the kind of detailed analysis of change through time that is possible in lake sequences, although in enclosed seas like the Mediterranean and in some other basins, sediment accumulation rates have often been faster than in most of the open ocean. Here, pollen and other analyses can help link deep-sea core records to those on land (Rossignol-Strick, 1995). At the opposite extreme, the stratigraphic record preserved in ice sheets can be thousands of metres thick. In the Summit cores from Greenland the Holocene is represented by no less than

Plate 2.5 Chemical sediments precipitated from a hyper-saline lake

1600 m vertical thickness of ice, and this permits annual time resolution – although coring to this depth does present some formidable logistical challenges (see Technical Box II, pp. 64–5).

Stable isotope analysis

As noted earlier in this chapter, most chemical elements have rare forms with the 'wrong' number of neutrons. Some of them, like ^{14}C, are isotopically unstable and undergo radioactive decay with time, but others are stable and provide information instead on past environmental conditions. Among the most important are the isotopes of carbon and oxygen.

The ratio between the heavy isotope oxygen-18 (^{18}O) and the lighter, and much more common, oxygen-16 (^{16}O) varies with temperature, and this fact led its discoverer – Harold Urey – to claim he had a 'geological thermometer' in his hands. Cesare Emiliani measured this $^{18}O/^{16}O$ (or $\partial^{18}O$) ratio in the fossil shells of tiny marine planktonic foraminifera to try to reconstruct changes in Pleistocene temperatures. The oxygen-isotope stages that he devised now form the way that the **Quaternary** ice age is classified into episodes of warmer and colder climate. It has since been established that a more important direct factor than temperature in influencing the marine isotopic record was the growth and decay of ice sheets.

These lock up isotopically 'light' water as ice, and leave the oceans enriched in the heavier isotope ^{18}O. In other words, oxygen-isotope analysis of deep-sea cores provides an index of past glaciation, rather than of temperature. Temperature has been a more important control over the $\partial^{18}O$ record in some other systems, such as freshwater lake sediments, ice sheets and speleothems. Oxygen isotopes can indicate seasonal as well as annual variations in temperature or rainfall, for example in mollusc shells as they accrete (Abell et al., 1995). In salt lakes, the stable oxygen-isotope ratio varies primarily according to the intensity of evaporation from the water surface, and here they reflect past salinity changes (Talbot, 1990).

Stable carbon-isotope analysis provides a way to determine past changes in the type of plants and in the global carbon cycle, for example as the atmospheric concentration of greenhouse gases has fluctuated. Some plants, such as tropical grasses and sedges, incorporate CO_2 as a four-carbon molecule during photosynthesis; the so-called C4 pathway. Other grasses, along with trees and shrubs, use a three-carbon (C3) pathway. The relative importance of C4 plants can be established by measuring the ratio of two stable isotopes of carbon, ^{13}C and ^{12}C. That ratio, expressed as $\partial^{13}C$, has been analysed in plant macrofossils, soils, ostracods, land snails, timbers and peats among others (Goodfriend, 1992; Street-Perrott et al., 1997; Heaton et al., 1995). In lake sediments, the stable carbon-isotope ratio will have been affected by the productivity of aquatic algae, as well as the type of vegetation in the lake catchment. Carbon isotopes have also been used to help determine past changes in human (and animal) diet. Maize, along with some other grain crops, incorporates CO_2 via the C4 pathway (Chisholm, 1989). The relative dietary importance of C4 plants can be established by measuring the $\partial^{13}C$ ratio in human bone. That ratio is high in the case of diets with a large proportion of C4 plants, such as maize, or of animals fed off C4 plants. The $\partial^{13}C$ values are low when C3 plants, including wild foodstuffs, dominate the food chain.

Geomorphology and climate

Climatic conditions exert a powerful influence over geomorphic processes and landforms. Geomorphologists such as Jean Tricart (1972) and Julius Büdel (1982) have argued that at a global scale distinct morphoclimatic regions can be recognized, within which a common set of processes are dominant. Thus ice-wedge polygons are exclusively found in the periglacial zone of winter freezing and summer thaw, where the average annual temperature is below −6°C (Black, 1976; Harris, 1986). Geomorphic processes such as freeze-thaw are so closely linked

Figure 2.11 The glacier–climate linkage

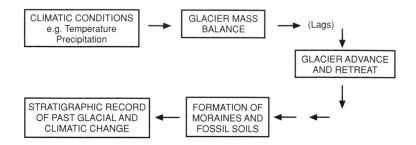

to climate that it is possible to calibrate them against temperature or precipitation. However, not all landforms are harmoniously adjusted to the prevailing climate because some were produced in the past when climatic conditions were different. These relict landforms may furnish indications about past or palaeoclimates.

One geomorphic regime intimately associated with specific climatic conditions is the GLACIAL environment. Valley glaciers form under cold, moist climates, specifically in areas where the accumulation of winter snow exceeds summer melting or ablation. The surplus snow builds up, becomes compressed to form ice, and moves downslope by basal sliding or flowage. In due course the glacier enters a zone of warmer air temperatures in which there is a net deficit between accumulation and ablation, and eventually the glacier terminates. The mass balance of a land-based glacier is therefore a function of winter snowfall on the one hand, and summer temperature and cloudiness on the other (Drewry, 1985). Under colder or more snowy conditions the glacier snout will advance downvalley, under warmer or drier conditions it will retreat. On the other hand, glacier response to a change in climate is rarely instantaneous, mainly because it takes time for an increase or decrease in ice volume in the accumulation zone to be felt at the snout (see figure 2.11). Even though changes are transmitted downvalley four times faster than the actual rate of ice movement, there will still be a time lag in glacial response amounting to a few decades for an alpine glacier, and a few millennia for a major ice sheet. In any case, portions of the major ice sheets – past and present – have had margins ending in the sea rather than on land, and in these cases sea level has been a more important control over ice-melt than air temperature.

Geologic evidence of former ice extent comes from moraines and till sequences which can be dated by ^{14}C measurement, dendrochronology and lichenometry. The existence of moraines several km downvalley from modern alpine glacier snouts represents convincing evidence of Holocene fluctuations in climate (Grove, 1979; Porter, 1981). The most recent of these moraines is little more than a century old and has only been

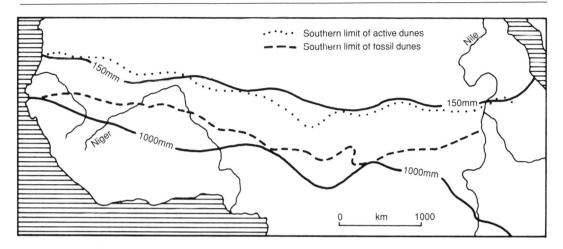

Figure 2.12 Modern and fossil dune limits on the south side of the Sahara (based on Goudie, 1983)

partially recolonized by vegetation (see plate 2.6). In contrast to those found in alpine valleys, the moraines created by Pleistocene ice sheets now lie far removed from any existing glaciers and are now covered by well-developed soils and vegetation. Some ice sheets have survived right through the Holocene, notably in Antarctica and Greenland, and cores drilled into them have provided high-resolution records of late Quaternary climate change (see Technical Box II, pp. 64–5).

Geomorphology provides important records of Holocene environmental change in the tropics and sub-tropics too, although in this case the primary climatic variations have been from wet to dry (Douglas and Spencer, 1985). Two of the most climatically sensitive geomorphic regimes are lakes and desert dunes. AEOLIAN processes are most effective where sediment is not bound together by surface soil or vegetation and which has a poor internal cohesiveness when dry. Wind-blown silt, or loess, originated in some cases from deflation of glacial sediments left behind by retreating ice sheets, and in others from high-altitude rock weathering in mountains. However, it is in arid lands where ideal conditions for aeolian processes are most often met, with climatically induced moisture deficiency causing vegetation to be sparse or absent. Because of the close relationship between aeolian processes and climatic aridity, fossil dunes help to provide an indication of the changing extent and location of the world's arid zone (Sarnthein, 1978; Lancaster, 1990).

On the southern (Sahelian) margin of the Sahara, active sand dunes are today restricted to land with less than 150 mm mean annual rainfall (see figure 2.12). Stable, vegetated dune ridges are none the less found up to 450 km south of this limit, in areas presently receiving up to 1000 mm rainfall (Grove and Warren, 1968). These mark a former extension of the Sahara southwards, when there was reduction of effective

Plate 2.6 Oblique air photograph of the Arolla (left), Pièce (centre) and Tsidjioure Nouve (right) glaciers in Switzerland. The moraines and trim-lines visible downvalley of the modern glaciers were created at times during the Holocene when the ice was more extensive and the climate cooler and snowier. The most recent advance occurred during the 'Little Ice Age', which ended in the nineteenth century

precipitation to only a quarter of its present value. Traces of fossil sand dunes have even been traced beneath the tropical rainforests of Zaire and Amazonia (Tricart, 1974). Aeolian landforms not only indicate times of climatic aridity but also reveal former wind strengths and directions (e.g. Wells, 1983; Bowler and Wasson, 1984; Thomas, 1984). Because dunes are aligned relative to the dominant wind vector, they may be used to establish palaeo-wind directions, and differences from the present can then be computed.

If dunes indicate periods of climatic aridity, high-level LAKES usually reflect wet phases or **pluvials**. In hydrologically closed or non-outlet lakes, water is lost only through evaporation, assuming there is negligible groundwater outflow (Street, 1980). Lake area and water depth adjust so as to increase or reduce evaporation losses, in the same way as a glacier expands or contracts in response to changes in snowfall and melting. In essence, closed lakes can behave like giant rain gauges: more rainfall (or less evaporation) and the water level rises, less rainfall (or more evaporation) and the water level falls. Direct geomorphological evidence of former high levels comes from shoreline terraces left 'high and dry' above present-day lakes; indirect evidence comes from proxy records of changing lake salinity (Street-Perrott and Harrison, 1985). When a lake moves from being open to closed, it also changes from freshwater to saline (Langbein, 1961). Instead of being washed away down the outflow stream, solutes are retained within the lake water and are progressively concentrated by evaporation. Some closed lakes have become so concentrated that they are now hyper-saline; the Dead Sea, for example, now has a salinity ten times that of ordinary sea water.

Palaeosalinities are recorded in the chemical composition of lake sediments, because different salts are precipitated as lake salinity increases; for instance, in a series from freshwater carbonates through sulphates to chlorides (Bowler, 1981; Teller and Last, 1990). Equally useful in salinity reconstruction are the surviving hard parts of aquatic organisms such as diatoms, ostracods and molluscs (de Deckker, 1981; Gasse, 1987; Holmes, 1992). Numerical estimates for past salinity and ionic composition can be obtained by means of statistical transfer functions which relate modern species assemblages to water chemistry in a 'training set' which spans a range of different environments. This approach has proved especially successful for diatoms (Fritz et al., 1991; Cumming and Smol, 1993). Ostracods can also be analysed chemically for trace elements such as strontium, calcium and magnesium, whose ratios are salinity- and temperature-dependent, and for stable isotopes (Chivas et al., 1986; Engstrom and Nelson, 1990).

Climatic calibration of lake-level data has been attempted using a number of different approaches. The first involves the calculation of past precipitation, runoff and evaporation for a lake and its catchment based on a model of the contemporary water balance (Street, 1980). The major assumption that this approach makes is that past temperatures – the primary control over evaporation rates – are known. An alternative approach developed by John Kutzbach (1980) utilizes a combined energy- and water-balance model. Both have been used to provide estimates of early Holocene precipitation in Africa and else-where in the tropics (Kutzbach, 1983). Some lakes are mainly fed from below rather than above, and act as groundwater 'windows' rather than as rain gauges. In such cases the whole groundwater system has to be modelled, as was done for the lakes of the Parker's Prairie Sandplain in the Mid-West United States (Almendinger, 1993).

Geo-archaeology

The interface between earth science and past human activity has been variously termed geo-archaeology (Davidson and

Shackley, 1976) and archaeological geology (Rapp and Gifford, 1982). It includes the study of sediments from archaeological sites but also extends off-site to investigate, for example, palaeo-geographic reconstructions of coastlines of historical importance (e.g. Kraft et al., 1977) and sites buried by river alluvium (Brown, 1997). One of the ways archaeology and geology have helped each other has been in the study of CAVE deposits. Most caves are ancient landforms, and during the Quaternary they have served as the residence of many creatures, including owls, bears and our own human ancestors. Indeed, it is popularly believed that all early hominids were 'cave men'. As caves have been infilled with sediment, so they have also incorporated in their stratigraphies materials brought in by their former occupants: bone remains, stone tools, charcoal from hearths and food refuse. These provide archaeologists with rich pickings for fieldwork, and cave sequences such as at Franchthi in Greece (Jacobsen and Farrand, 1987) and Cresswell Crags in England (Jenkinson, 1984) have been meticulously excavated in recent years in collaboration with environmental scientists, often employing techniques such as micromorphological analysis (Courty et al., 1989).

Interdisciplinary research of the geo-archaeological type has many potential advantages (Vita-Finzi, 1978; Stein and Farrand, 1985). Archaeological and palaeoenvironmental data can be matched and dated from the same stratigraphic context without the need for correlation between sites. In the case of Lake Mungo in Australia, it was survey work on lunette dunes by geomorphologist Jim Bowler that led to discovery of an archaeological site in the first place (Bowler et al., 1972). The human remains found at Mungo pushed back the antiquity of aboriginal peoples in Australia to over 25 000 Cal. yr BP, and also helped to provide a timescale for the geological and climatic changes recorded in the site's sediment stratigraphy (see figure 2.13). In particular the Mungo skeleton provided an age for the beach gravels with which it was associated, and hence for a phase of high lake levels.

Over the course of the Holocene, human impact on geomorphic processes has increased progressively, especially on the rate of SOIL LOSS from slopes. Soil is eroded and lost as part of the natural geological cycle of denudation, but under dense grassland or woodland that loss is minimized because rainfall is intercepted and runoff reduced (Thornes, 1987). Removal of the vegetation, by human or natural agencies, causes increased potential for erosion by raindrop impact, surface sheetwash, rilling, gullying, leaching and wind action (Limbrey, 1975; Boardman et al., 1990). Nor is impact restricted to the site of degradation. Eroded soils are washed downslope and downstream to push up the sediment loads carried by rivers.

In fact, soil erosion is an important indicator of the degree of disturbance in an ecosystem.

Geomorphic environments which record soil erosion histories may be broadly divided into those of net erosion and those of net deposition (Bell, 1983; Bell and Boardman, 1992; Dearing, 1994). The former ought to provide the more accurate soil loss data, but on their own, degraded hillslopes are an unreliable guide to erosion history, their mere existence providing no clue to the age or origin of erosion. However, where archaeological sites lie *in situ*, the former soil surface may be preserved beneath them. Examination of soils beneath prehistoric burial mounds has shown that on the English chalk downlands, later Holocene erosion has almost completely removed the original cover of loess (Catt, 1978).

Most studies of past erosion rates have focused on the point of deposition rather than the point of erosion. Suitable environments include valley fills, floodplain and estuarine deposits, and lake sediments. In all cases sediment yield is used as a surrogate for soil loss, which tends to underestimate true erosion rates. In using sedimentary records it is best to use relatively small, 'closed' catchments such as drainage basins with a high trap efficiency (e.g. Brown and Barber, 1985), often containing a lake or reservoir (e.g. Foster et al., 1985). Mineral magnetic properties such as **magnetic susceptibility** and magnetic remanence allow different types of iron minerals to be distinguished, and this often helps pinpoint the source area from which sediments were derived (Thompson and Oldfield, 1986). Soils, for example, have a different magnetic 'signature' from the parent bedrock because of the formation of secondary ferrimagnetic oxides such as magnetite, and this is enhanced if the soil has been subject to burning.

Models of Environmental Reconstruction

Up to now, palaeoenvironmental techniques have been dealt with individually. The question therefore remains of how they and individual site investigations should be fitted together to provide a comprehensive picture of landscape change. One conceptual framework frequently used is the environmental system model, in which different physical, chemical and biological components of the environment are seen to be interconnected over a definable area of space. A good example is the LAKE-CATCHMENT ECOSYSTEM, in which a natural and easily defined system boundary is provided by the drainage basin watershed (Oldfield, 1977; O'Sullivan, 1979). Nutrients and other materials are washed down slopes into streams, and eventually end up in the lake. Not only does the lake therefore

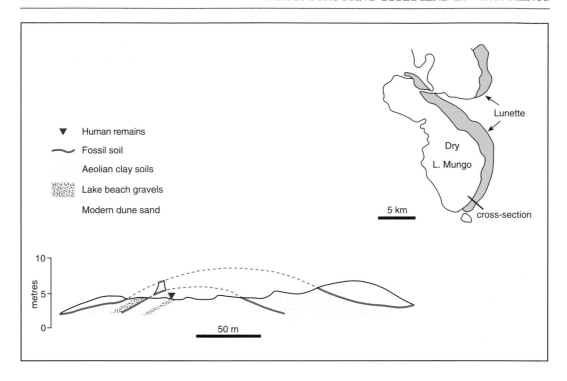

Figure 2.13
Sediment and
archaeological
sequence at Lake
Mungo, Australia
(after Bowler et al.,
1972)

serve to integrate changes over the whole catchment – that is, across space – it also records the history of such changes through time. Whereas in a normal drainage basin materials would be lost from the environmental system along with stream discharge, in a lake catchment they are trapped and incorporated in bottom sediments, so preserving a record of past environmental processes (Binford et al., 1983; Smol, 1992). Furthermore, different techniques of dating and analysis may be used in combination on the same lake sediment cores; for example, to reconstruct both catchment erosion rates via sediment influx calculations, and a history of land use via the pollen record.

So although palaeoenvironmental data typically derive from specific field localities, the material they contain does not originate solely from that one place. In other words, each individual site locus actually represents an aggregated record over a wider catchment area. The size and shape of each site catchment will vary according to the type of site and other factors (Vita-Finzi, 1974). Whereas a lake's catchment is effectively the same as its drainage basin, that of a pollen core on a peat bog is defined atmospherically, being the area upwind during the pollen production season (see figure 2.14). Birds, animals and human populations may be thought of as having catchments too. The faunal remains found in a cave reflect the area exploited by its occupants, such as the hunting territories of

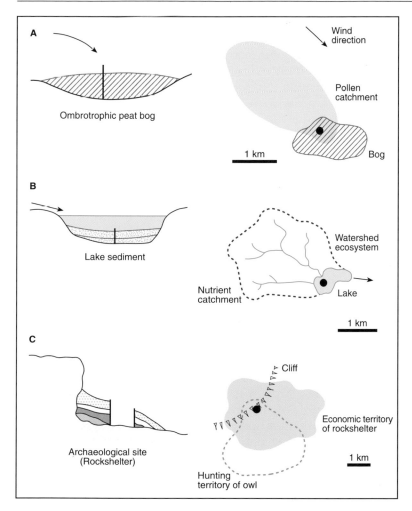

Figure 2.14
Palaeoenvironmental sites and their catchments: **A** ombrotrophic peat bog; **B** lake watershed ecosystem; **C** rockshelter

an owl or a Mesolithic hunting band. The nature of the territory around an archaeological site helps to explain the on-site record of economic activities preserved in seed and bone remains, or in historical archives for individual estates or monasteries (Higgs and Vita-Finzi, 1972; Butlin and Roberts, 1995).

Of course, catchment areas may not be so easily defined as the examples shown in figure 2.14. Even so, they serve to emphasize that environmental reconstruction takes place principally at the landscape scale (Birks et al., 1988). In order to study environmental processes operating at a smaller scale – say, within a stand of trees – it is necessary to select those sites which only reflect local changes (Bradshaw, 1988). Whereas the pollen entering a large lake normally reflects regional-scale vegetation (c.100 km²), that found in a small woodland hollow will have travelled a short distance and have been determined instead by the nearby flora (c.0.1 ha). At the opposite extreme, large-scale changes involving vegetation formations or climatic zones (c.1 million km^2) can only be derived by combining data from many individual sites. For example, pollen records from over 400 core localities are held in the North American Pollen Database, and 650 in its equivalent for Europe, both used to reconstruct continent-wide vegetation changes since the last glacial maximum.

Just as there is a spatial resolution to the reconstruction of past environments, so there is also a temporal resolution. Palaeoecological and geological evidence is resolvable down to one year in the case of varves and tree-rings, but more usually the time interval between sample points is 10 to 100 years (see table 2.2). While this inevitably means the 'loss' of information about year-to-year environmental variations, it is compensated by the 'gain' of a smoothed record of underlying trends over centuries and millennia. These are timescales of change which are too long to be observed directly, and for which proxy records are uniquely well suited.

Tying the various complementary sources of palaeo-environmental data together needs good liaison between different specialists. For this reason many projects on environmental history involve a team of researchers working over a number of years at a particular site or in a particular area, whether this be on a long core through an ice cap, or a study of cultural transformation of a regional landscape. Integrating information from archaeological excavations, historical records, off-site pollen cores and alluvial sequences, and dating them all, can be a major challenge (not to say headache), but at its best the combined results can reveal much more than the sum of their component parts. For example, in the Ystad project a coordinated team of archaeologists, palaeoecologists, landscape historians and others worked together over a number of years to uncover the history of one area of southern Sweden (Berglund, 1991). There are limitations to any one technique, but together they are able to provide us with an insight into a past that would otherwise remain inaccessible.

Only selected techniques of dating and environmental reconstruction have been reviewed in this chapter or included in subsequent technical boxes. Fortunately, several texts provide further detail on these and other palaeoenvironmental methods, for example, by Birks and Birks (1980), Bradley (1984), Berglund (1986), and Lowe and Walker (1997).

CHAPTER THREE

The Pleistocene Prelude (>11 500 Cal. yr BP)

Ice-age Environments

If we draw back in time from the start of the Holocene, some 11 500 years ago, it soon becomes clear that our present epoch is but the latest of a series of warmer intervals that have punctuated the otherwise chilly climate of the Quaternary. The last few million years have been marked by frequent climatic changes, with repeated oscillations between colder and warmer conditions. Warmer intervals lasting more than about 10 000 years are known as interglacials, while the colder periods are termed glacials. During the latter, massive ice sheets developed over Scandinavia and Canada, only to melt away entirely during the interglacials. Interspersed within glacials and interglacials were cold and warm intervals of shorter duration (e.g. 1000 years), known as **stadials** and **interstadials**, respectively.

The glacial–interglacial cycle

For many years a scheme involving four Pleistocene glacial–interglacial cycles was recognized or even imposed in many land areas. The evidence for this view of the Quaternary ice age originally came from moraines, erratic boulders, striated and polished rocks in areas such as the alpine foreland of Switzerland and southern Germany. Since 1950, more complete records of Pleistocene climatic change have been available from the study of deep-sea sediment cores, in particular using oxygen-isotope analysis. Deep-sea cores provide unbroken stratigraphic records and have shown that schemes like the four-glacial model established by Penck and Bruckner for the Alps have underestimated the true number. About eight major glacial–interglacial cycles have in fact occurred during the past 0.8 million years, each cycle lasting c.100 000 years, and with others before this time (Williams et al., 1993). The deep-sea record of past climate change now has its counterpart on land from thick loess deposits in central Europe and Asia, and long pollen records from sites in Greece, the Colombian Andes and elsewhere (Hooghiemstra et al., 1993; Porter and Zhisheng, 1995; Tzedakis, 1993).

Evidence for interglacial conditions comes from fossil soils, buried peat and lake sediments, and other deposits containing warm-climate faunas. The last true interglacial, known in continental Europe as the Eemian, in Britain as the Ipswichian and by other names elsewhere, began a little after 130 000 years ago, and lasted for around 11 000 years. During that time, it appears that climatic conditions were close to those which have been experienced during the Holocene. Temperatures were within 2°C and sea levels within a few metres of those during recent millennia (CLIMAP project members, 1984), while sub-tropical areas such as the Sahara experienced the same increase in rainfall recorded during the first half of the Holocene (Gaven et al., 1981). A recent ice-core record from Greenland has suggested that the last interglacial climate may have been more unstable than that of the Holocene, and cold episodes have also been inferred from within some Eemian pollen sequences (Field et al., 1994; Tzedakis et al., 1994). On the other hand, these short-lived stadials were not found in a second Greenland ice core (see Technical Box II, pp. 64–5) and the interpretation of the pollen results has not met with univeral agreement (Aaby and Tauber, 1995). Earlier interglacials also seem to have been very similar in character to the Holocene, although counting of varved lake sediments indicates that some, such as the British Hoxnian, may have lasted longer – perhaps for 20 000 years (Turner, 1970).

In one respect, however, the Holocene stands in sharp contrast with the Pleistocene interglacials – the mark of 'man' imposes itself far more dramatically on the present warm epoch than on any previous one. The emergence and evolution of the human race is one of the most important distinguishing characteristics of the Quaternary, but anatomically modern humans (*Homo sapiens* var. *sapiens*) only emerged 100–200 000 years ago. The main radiation took place during the last glacial period, with Neanderthal populations being replaced in western Europe about 40 000 years ago (Lewin, 1993). During earlier Pleistocene interglacials, therefore, only ancestral humans were present, and in some parts of the globe, such as the Americas, even they were absent.

Given that the impact of early hominids on the natural world was rather slight, it is possible to use Pleistocene interglacials as analogues for what the Holocene world would have been like had it not been disturbed by human agency. Pollen analysis of peat and lake deposits has revealed the vegetation changes that took place during Pleistocene interglacial periods (Davis, 1976a). Although there were some floristic differences between interglacials, they were minor in comparison with their overall similarity. This is all the more remarkable considering that many plant formations had to be recreated from scratch at the beginning of each warm interval. In northern Europe, Johannes Iversen (1958) proposed a simple phase model for interglacial vegetation development. The pre-temperate **protocratic** phase was one of immigration of tree species from southern refuges, and the first arrivals were normally the boreal taxa of birch (*Betula*) and pine (*Pinus*). During the subsequent **mesocratic** phase, mixed deciduous forest became established (see table 3.1). Shade-giving trees, such as oak (*Quercus*) and elm (*Ulmus*), replaced the pioneering light-demanding genera. Although the mesocratic phase was usually of longer duration than the

Table 3.1 Ecological characteristics of trees associated with different phases of the northern European interglacial cycle

Ecological characteristic	Protocratic	Mesocratic	Oligocratic
typical tree taxa	birch	oak	beech
	aspen	elm	spruce
age of first seed setting	young	mature	mature
seedling tolerance to shade	intolerant	tolerant	tolerant
dispersal efficiency	good	poor	poor
migration rate (m/yr)	>1000	500–1000	<500
growth rate	fast	slow	slow
longevity	short	long	long
soil preference	fertile	brown earth	podsol
	unleached	mull humus	mor humus
ecological traits	ruderal	competitive	stress-tolerant

Source: Birks (1986a)

protocratic one, it was far from static floristically. Pollen diagrams show that new trees continued to arrive throughout the duration of interglacials, with floristic diversity progressively increasing as a result (Davis, 1976a).

Soil changes also influenced the composition of the temperate forests, especially in areas that had been subject to glaciation during cold periods of the Pleistocene (see figure 3.1). Here, leaching of weathered bases from glacial tills led to a shift from neutral to acid soils, favouring trees like spruce (*Picea*), which mark the **oligocratic** phase of an interglacial. Once established, conifers produced further soil acidification as their fallen needles created an acid mor humus (Andersen, 1969). The final retrogressive **telocratic** phase saw the replacement of mixed deciduous forest by heathland and open coniferous woods, particularly where soils were locally prone to leaching out of their nutrients. Thermophilous trees disappeared as the climate changed from interglacial to glacial, but the order of departure was not simply that of arrival in reverse. It seems as if many deciduous trees were not even able to 'pack their trunks' and to head off towards the Mediterranean sun at the end of the last interglacial (Woillard, 1979; Bennett et al., 1991).

Most high-latitude pollen records stop with the switch from interglacial to glacial conditions. On the other hand, this is not so with sites from the tropics and sub-tropics. They show that vegetation formations such as the rainforest of northern Australia and the xerophytic woodlands of the Mediterranean basin were virtually eliminated during glacial phases of the Pleistocene (Kershaw, 1978; Tzedakis, 1993). The tropical rainforests of Africa and Amazonia were also reduced to isolated

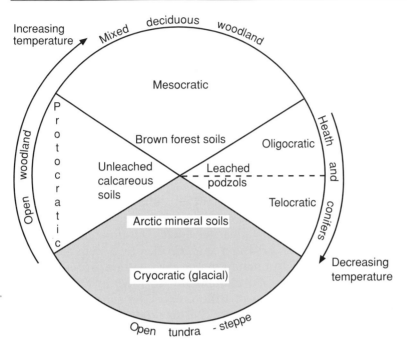

Figure 3.1 The interglacial cycle in northern Europe, e.g. Denmark (modified from Iversen, 1958)

pockets at these times. Climatically induced vegetation change was consequently not restricted to the northern forests, but affected many of the world's biomes. What is more, this great natural deforestation of the Earth was repeated many times during the course of the Pleistocene.

In short, the million years or so prior to the Holocene was a period of extraordinary environmental variability and stress upon ecosystems and organisms. Those species capable of adapting to a fluctuating environment were able to benefit. Our own species is a prime example, for Pleistocene environmental change undoubtedly helped speed the pace of human evolution (Roberts, 1984). There were unusually high rates of speciation and extinction in many areas of the world during the Pleistocene. In South America, for instance, repeated fragmentation and coalescence of the rainforest encouraged the creation of new species and may account for the high species diversity of that biome (Prance, 1982; Colinvaux, 1987). In most cases, however, evolutionary changes brought about by Pleistocene fluctuations in climate were secondary to those which occurred in species distribution and population numbers. During the Pleistocene, species shifted their geographic distributions, many experiencing alternate fragmentation and reconstitution of their ranges, while others simply tracked their habitats by moving towards or away from the poles, or up and down mountains. The migratory shifts frequently involved distances of several thousands of km in organisms ranging from plants and insects to mammals (Bennett, 1997).

Soil- and land-forming processes were subject to a similar intensity and rapidity of change during glacial–interglacial cycles. This was obviously most acute in areas which were glaciated. The soils of the Canadian Shield were literally scraped up and dumped in the American Mid-West every time the Laurentide ice sheet advanced southwards – a later gift from Canada to America's farmers.

Understanding the causes of long-term climatic change

It is now widely accepted that the main trigger for ice-age climatic fluctuations was small variations in the Earth's orbit around the sun. This idea goes back to nineteenth-century scientists such as Adhémar and Croll, but a coherent astronomical theory of climate change was not put together until the 1920s. The Serbian mathematician Milutin Milankovitch identified three cycles of different length (table 3.2), which were superimposed on each other to produce a more complex pattern of change through time (Berger, 1992). Only one of these, the eccentricity cycle, in fact caused the Earth as a whole to receive different amounts of solar radiation. The other two factors (tilt and precession) involved its redistribution between seasons in the different hemispheres. On that basis we might have expected glacial phases to have taken place alternately in the northern and the southern hemispheres. In reality, the pattern of glacial and interglacial stages was the same all over the globe. It seems, therefore, that the effects of orbital variations over the northern hemisphere land masses triggered mechanisms which acted to cool or warm the planet as a whole. Comparison of Milankovitch's predictions with the deep-sea record of past ice-sheet expansion and contraction has shown a remarkably close fit – prompting claims that the mystery of ice-age changes in climate has been solved (Imbrie and Imbrie, 1979). Certainly the astronomical theory has proved invaluable, not least because it allows us to make predictions about the future course of climate. It is clear, for example, that we are in the latter part of an interglacial period, and that if allowed to run its course naturally the Earth's climate would be back in a glacial phase some 25 000 years from now.

Table 3.2 The main astronomical cycles of climatic change

	Cycle length (years)
Orbital eccentricity	100 000
Axial tilt (or obliquity)	41 000
Precession of the equinoxes	21 000

On the other hand, changes in the Earth's receipt of solar radiation due to orbital variations were relatively modest – only sufficient to cause direct warming or cooling of around 2°C. In reality, temperatures fluctuated by much more than this amount in the past, especially on land. For this reason, Jim Hays and colleagues (1976) aptly described variations in the Earth's orbit as the 'pacemaker' of the ice-age climate. Other forces must have been at work to amplify its signal, some of which would have been internal to the Earth's own system. One important set of feedbacks involves the linkages between ice, ocean and climate. Many past or present ice sheets, for example, terminate in the oceans, and sea-water temperature and sea level, rather than local climate, control how far out the ice is able to grow. A minor increase in melting of the ice triggered by orbital changes would raise sea levels. This in turn would undercut more ice (often by destabilizing the floating ice shelves that protect the ice-sheet margins from wave attack), and the ice calved off into the sea would raise sea levels still further. This self-reinforcing process would continue until the ice retreated to the point where it was grounded on land, not in the ocean, or until some other factors came into operation to stop the positive feedback loop. Other feedback links between ice, ocean and climate involve surface albedo effects, and the impact of cold, fresh meltwater on ocean circulation and on evaporation rates from the ocean surface. In these ways, a small externally driven initial warming could have been amplified to push the climate from a glacial to an interglacial state, or vice versa. Indeed, the Earth's ice-age climate appears to have alternated between two more or less stable states (Broecker et al., 1985).

Another process which may have amplified the orbital climatic signal was changes in greenhouse-gas concentrations. Gases such as carbon dioxide and methane help 'trap' heat within the atmosphere by intercepting outgoing long-wave radiation, and higher atmospheric concentrations would have been associated with enhanced warming of the climate. Estimates of past fluctuations in greenhouse-gas concentrations come from sources such as the stable carbon isotope ($\partial^{13}C$) ratios in peats, lake sediments and soils (Street-Perrott et al., 1997; Figge and White, 1995). However, our most informative source of information comes from ancient air bubbles trapped in ice sheets and recovered in ice cores (see Technical Box II). The long Vostok core from Antarctica provides our most complete and direct record of changes in greenhouse gases and temperatures, spanning the whole of the last glacial–interglacial cycle. As figure 3.2 shows, during interglacials CO_2 levels were around 270–280 ppmv, while under full glacial conditions they fell to 190–200 ppmv. The close match between this curve and that

Figure 3.2 Variations in carbon dioxide and temperature through the last full glacial–interglacial cycle in the Vostok core from Antarctica (after Lorius et al., 1990)

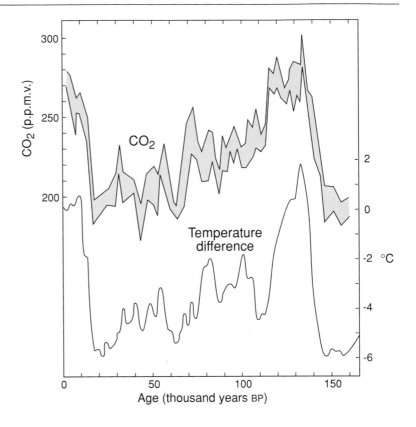

for temperature strongly supports a link between greenhouse gases and past climate change; the former appearing to act as an amplifier of the Milankovitch astronomical signal.

The last glacial maximum and after

The last glacial period reached a peak between 25 000 and 18 000 years ago according to calibrated radiocarbon dating. Evidence for what the world was like at this time is relatively abundant and well preserved, certainly in comparison with earlier glacial periods whose record has been greatly disturbed by later ice action and other geomorphic processes. Around 21 000 Cal. yr BP, temperatures on land had fallen by as much as 20°C (Wright et al., 1993), although ocean temperatures changed much less. The reconstruction by the CLIMAP project members (1976) indicated that for tropical parts of the ocean, sea-surface temperatures for the last glacial maximum were within 2°C of the present. More recent work has suggested that the true temperature fall may have been nearer to 5°C, however (Guilderson et al., 1994). Under the full glacial climate, ice sheets over 4 km thick lay over northern Europe and North America, and smaller ice caps existed in the Alps, the south-

	LGM	Holocene
Tropical forest	11.9	26.1
Temperate forest	7.1	33.7
(Total forest)	(19.0)	(59.8)
Savanna and scrub	44.0	31.1
Temperate grassland/steppe	11.3	15.5
Desert	29.5	17.2
Tundra and polar desert	27.1	9.0
(Total non-forest)	(111.9)	(72.8)
Other (mainly ice)	33.4	17.4
Total	165.3	150.1

Table 3.3 The extent of major terrestrial biomes at the last glacial maximum (LGM) and the late Holocene (excluding agricultural clearance) (areas in $\times 10^6$ km^2)

Source: Adams et al. (1990)

ern Andes and parts of East Asia (Denton and Hughes, 1981). Most of northern Asia was too dry for ice sheets to form, and Siberia consequently experienced intense periglacial activity. Much of today's deep permafrost layer is, in fact, a relict from the last glacial period.

The build-up and decay of ice sheets locked up and then released water from the hydrological cycle, causing ocean volume to fluctuate and sea levels to rise and fall (Tooley and Shennan, 1987). Global sea levels were lowered on average by more than 100 m at glacial maxima, creating new continental land masses in areas such as southeast Asia. Land bridges were exposed, so that it became possible to walk from east Asia to Alaska or from Britain to the European mainland. This undoubtedly aided the spread of fauna and of early human populations into new or marginal areas. It was the era of the woolly mammoth, the American sabre-tooth tiger and of the ice-age hunters who portrayed their prey so vividly on the cave walls at Lascaux in southwest France.

Most of northern Europe that was not glaciated formed a tundra-like landscape at this time. Some shrubs such as dwarf birch (*Betula nana*) may have survived the harsh climate, but vegetation was sparse and was often dominated by steppic plants. Open-habitat species predominated in most of southern Europe as well, and trees were restricted to isolated refuges in Iberia and southeast Europe (Bennett et al., 1991). On the other hand, this was not true of all mid-latitude regions. Japan almost certainly remained forested through the last glaciation (Tsukada, 1983), while much of the eastern United States was covered by pine and spruce woodland (Webb et al., 1993). A tundra zone did exist close to the Laurentide ice sheet, but in contrast to northern Europe, it was only 100–200 km broad (Watts, 1983).

Technical Box II: Ice cores and climate change

Cores drilled through the layered accumulations of snow and ice deposited in the polar or high-mountain ice sheets permit a detailed reconstruction of past climate (plate 3.1). The longest record, from the Russian core site at Vostok in Antarctica, extends back through several full glacial–interglacial cycles (Jouzel et al., 1993). The ratios of hydrogen and oxygen isotopes in the ice provide an index of former temperatures, while the dust content indicates global wind strength and aridity, and its acidity provides an index of major volcanic eruptions (Zielinski et al., 1994). Furthermore, snow, as it accumulates and is compressed into ice, traps air bubbles whose methane (CH_4) and carbon dioxide (CO_2) contents can be measured from the cores to provide a record of the changing concentrations of these greenhouse gases (Lorius et al., 1990). This applies not only to long timescales when the climate varied due to natural factors, but also to historic changes associated with fossil fuel burning and other human impacts.

Plate 3.1 Annual snow accumulation layers, thicker at the surface and compressed with depth, are clearly visible in the edge of this Antarctic ice sheet. These annual bands provide a ready-made timescale for ice-core records of past climatic change

Ice does not only build up vertically, however; it also flows out laterally from the centre of the ice sheet to its margins. For this reason, cores taken at the top of the ice-sheet dome should have experienced less deformation and contain a longer record than those taken near the edges. On the other hand,

TECHNICAL BOX

the high points of ice sheets are hardly the most accessible or hospitable places on Earth; Vostok has a mean annual temperature of −55°C and stands 3490 m above sea level. In the case of the Huarascán ice cap on the top of the Peruvian Andes, Lonnie Thompson and his team had to haul six tonnes of solar-powered drilling equipment manually up to over 6000 m altitude (Thompson et al., 1995)! His latest plan in Bolivia involves bringing ice cores down by hot-air balloon . . .

In the northern hemisphere, Greenland contains the most important record. Ice has accumulated faster here than in Antarctica and it therefore offers a more detailed sequence of past climatic change. In a competition reminiscent of Amundson and Scott's race to the South Pole, two separate teams – one European, the other American – drilled cores only 28 km apart at the summit of the Greenland ice sheet between 1988 and 1993. (The Europeans 'won' when they reached bedrock at 3028 m on 12 July 1992, just under a year ahead of the Americans.) The apparent duplication of effort in producing two ice cores, called GRIP and GISP2 respectively, allowed a cross-check to be made, which fortunately revealed excellent agreement, except for the last interglacial (Boulton, 1993; Taylor et al., 1993). The two Greenland Summit cores have provided a very important climate record, which has an annual time resolution for the part covering the late Pleistocene and Holocene.

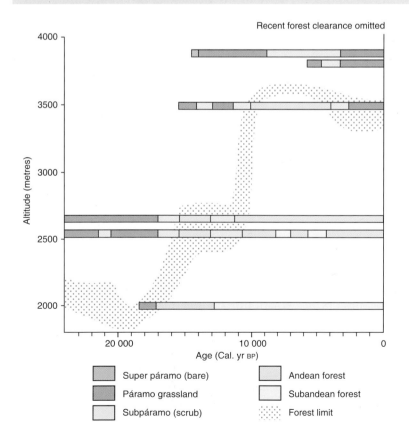

Figure 3.3 Altitudinal shifts of vegetation belts in the Colombian Andes during the late Quaternary, inferred from six pollen sites (modified from Flenley, 1979b)

In the tropics, temperature depression caused vegetation zones on the equatorial mountains to be lowered by 1000–2000 m (see figure 3.3), but a bigger change was brought about by glacial aridity (Goudie, 1983; Bonnefille et al., 1990). During and after the last glacial maximum, most low-latitude areas were significantly drier than at present. This is indicated by geomorphic data, notably low lake levels (Street-Perrott et al., 1985) and fossil sand dunes (Sarnthein, 1978; Stokes et al., 1997). The latter extended way beyond their present isohyetal limits and in some areas even underlie modern rainforest (see figure 2.12, p. 47). The Indian monsoon was suppressed, reducing the transfer of water vapour from the oceans to the continents. However, there were exceptions to this widespread aridity, including the rainforest of southeast Asia, which seems to have survived intact through glacial times. Another relatively well-watered area was the American Southwest, where a complex system of cascading lakes occupied the now parched deserts of the Owens River and Death Valley prior to 11 000 years ago (Smith and Street-Perrott, 1983). Packrat midden studies have revealed that conifer woodland then occupied what is today desert scrub (figure 2.9, p. 40).

The cold and general aridity of glacial times was not conducive to plant growth, particularly of trees, and led to expansion of savanna grassland, tundra, steppe and other open types of vegetation at the expense of forest biomes (table 3.3). This in turn led to significantly less carbon being stored in land vegetation and soils. One study (Adams et al., 1990) estimated that this organic carbon store was less than half that in the modern 'natural' biosphere, but other calculations indicate that the true reduction at the last glacial maximum was more likely to have been around 500×10^{15} g, a decrease of about one-fifth (van Campo et al., 1993; Prentice et al., 1993). This presents us with something of a conundrum. Ice cores show that there was about 30 per cent less carbon in the glacial atmosphere, but far from being taken up by land plants and soils, this carbon store too was depleted. So where did the 'missing' carbon go in the ice-age Earth? The answer would appear to lie in the oceans. Some of the removal of carbon can be accounted for by more efficient 'export' of organic matter from the upper photic layer of the oceans to the sea bed, where it is buried in sediments. But most of it appears to have been caused by a real increase in the productivity of marine plankton.

Most open oceans are biological deserts, their low productivity being caused by a shortage of key nutrients such as phosphorus and nitrogen. In many areas of the ocean, however, the nutrient which limits plankton growth is iron (Martin and Fitzwater, 1988), notwithstanding the fact that this is one of the commonest elements of the Earth's lithosphere! It both controls biological productivity directly and also helps to regulate the amount of nitrogen available to plankton (Falkowski, 1997). Iron reaches the open ocean mainly in the form of wind-blown dust, carried from the continents. The flux of atmospheric dust depends on conditions in dryland source areas, and during glacial times it was a good deal drier, and windier too. This is evident in the record of lake levels, loess and sand dunes in key source areas such as Patagonia, which lies upwind of the vast Southern Ocean; it is also recorded in greatly elevated dust concentrations in the Greenland and Antarctic ice cores (see figure 3.5C). The iron limitation on productivity in many open ocean areas therefore appears to have been relieved in glacial periods by a more than fivefold increase in the dust supply carried from the continents. Ironically therefore, the same arid climatic conditions which *decreased* biomass on land appear to have *increased* it in the oceans, and in so doing they contributed to lowering the concentration of carbon dioxide in the glacial atmosphere (Kumar et al., 1995).

After about 18 000 Cal. yr BP, the world's major ice sheets retreated and thinned. Deglaciation, which took about 9000 years to complete in Europe and North America, may have had as much to do with climatic aridity as with warming. The very bulk of the ice sheets created anticyclonic conditions and deflected moisture-bearing winds away from the ice sheets' accumulation zone (Wells, 1983). Evaporation from the ocean surface was also discouraged by a cool, low salinity meltwater layer, notably in the North Atlantic (Ruddiman and McIntyre, 1981). The ice sheets therefore appear to have been starved to death by inadequate nourishment of snowfall. As the ice melted and sea levels started to rise, so the ice sheets became vulnerable to attack by marine action, especially in areas such as Baffin Sound. Great blocks were undercut by the waves at the ice-sheet margins, to float away as icebergs and eventually raise sea levels still further. The southern, land-based margin of the Laurentide ice sheet retreated more slowly during the early stages of deglaciation, and in consequence the ice became thinner rather than smaller in extent.

Deglaciation was an unsteady process, and at various times towards the tail end of the Pleistocene the ice sheets temporarily readvanced. These glacial readvances do not appear to have occurred synchronously everywhere. The Laurentide ice sheet even expanded on some sectors of its southern front while other parts were stationary or retreating (Dyke and Prest, 1987). The Green Bay, Lake Michigan and other ice lobes of the Great Lakes region may owe their origin to glacial surging rather than to climatic forcing. Indeed, there is evidence of major, concentrated iceberg discharges from the northern and eastern edge of the Laurentide ice sheet into the North Atlantic on half a dozen occasions during the last glacial stage – the so-called 'Heinrich events' (Broecker, 1994). These seem to have occurred when the ice sheet went from a state of being frozen to the bedrock, to one in which the bed was above pressure-melting point. On a lubricated bed, the ice thinned and surged outwards, much of it into the sea to create a vast armada of icebergs. However, in parts of the world other than North America, environmental changes occurred during the terminal Pleistocene that were undoubtedly caused by a sharp oscillation in climate, which represented a false start to the present (Holocene) interglacial.

The Terminal Pleistocene (16 000–11 500 Cal. yr BP)

The Late-glacial in the British Isles

In Britain and Ireland the late Pleistocene thermal oscillation comprised an interstadial (15 000–13 000 Cal. yr BP) followed by a stadial (13 000–11 500 Cal. yr BP). Some of the best evidence for this oscillation comes from the remains of invertebrates, especially beetles and chironomids (see Technical Box III), which provide the next best thing to a Late-glacial thermometer. After rising abruptly around 15 000 Cal. yr BP, summer temperatures lay at – or even above – those of today, but Coleoptera suggest the climate then cooled progressively from the optimum of this 'Windermere' interstadial (Lowe et al., 1995), equivalent to the Bølling–Allerød events of Scandinavia (table 3.4).

Vegetation was slower than insects in responding to the climatic improvement at the beginning of the interstadial. Plants require suitable soil conditions and prior to 15 000 Cal. yr BP soils were thin and undeveloped, although a widespread change in lake sediment stratigraphies from inorganic clays to organic muds indicates that soil development and inwash occurred soon thereafter. The initial beneficiaries of warmer conditions were open-habitat species of low competitive ability, such as grasses and sedges. Evidence from absolute pollen counts of a sharp rise of pollen influx (Pennington and Bonny, 1970) indicates that this open vegetation responded by increasing its density and biomass. Changes in species composition occurred soon afterwards, as secondary colonizers in the plant succession invaded. These were, for the most part, shrubs such as juniper (*Juniperus*), willow (*Salix*) and crowberry (*Empetrum*). Open or scrub vegetation was to dominate in Ireland, Scotland and northeast England for the remainder of the interstadial

Technical Box III: Insect analysis

Among the beetles (or Coleoptera) alone there are about
350 000 different species – more than all the flowering plants
combined – so, not surprisingly, many are adapted to highly
specific ecological niches. For example, many species have
fastidious preferences for warmth or cold. Insect taxonomy
is based on the characteristics of their exo-skeletons. Skeletons are often well preserved in sediments but become
disaggregated into their component parts (head, thorax, etc.)
(see plate 3.2).

Plate 3.2 Modern and fossil specimens of beetle genitalia, *c*.0.8
mm in length, of the genus *Helophorus*. In some cases the size
of the male sexual organ is the only way to distinguish between
cold-loving Siberian species from temperate European ones;
remarkably, genitalia of Late-glacial age have been found
preserved

To date, the main contribution has come from fossil
Coleoptera, in particular in the reconstruction of late
Pleistocene palaeoclimates in mid-to-high latitude regions
such as northwest Europe. Their thermal likes and dislikes,
coupled with an ability to migrate rapidly, make Coleoptera
ideal indicators of past temperatures. Studies at a variety of
European sites have allowed Russell Coope (1975; Coope
and Lemdahl, 1995) to estimate average July temperature
changes from before 15 000 to about 10 500 Cal. yr BP. At
the beginning of this period temperate assemblages replaced
arctic ones, and at Glanllynnau in north Wales the rate of
climatic amelioration is calculated to have been – at least –

TECHNICAL BOX

an amazingly rapid 1°C per decade. Beetle assemblages can now be climatically calibrated for both summer and winter temperatures using the mutual climatic range method (Atkinson et al., 1987). Holocene climatic reconstructions based on Coleoptera have so far proved more problematic.

Insects which have to spend the early part of their life cycle living in water, such as Trichoptera (caddisflies) and chironomids (midge larvae), are also proving to be valuable sources of information about past environmental conditions (Walker et al., 1991b; Williams, 1988). Chironomids are proving to be just as sensitive to warmth and cold as beetles, and their preserved head capsules have been used to reconstruct Late-glacial temperature fluctuations in eastern Canada and northwest Europe (Walker et al., 1991a, b; Brookes et al., 1997). They have also provided an index of changing lake-water quality later in the Holocene (Carter, 1977; Sadler and Jones, 1997).

(Huntley and Birks, 1983), but in most of England and Wales it was succeeded towards 14 000 Cal. yr BP by birch woodland, with downy birch (*Betula pubescens*) being the most important species. Some British and Irish sites suggest a brief cold snap comparable to the Danish–Scandinavian Older Dryas around 14 000 years ago, but the consequences of this mini-stadial for vegetation were short-lived and localized.

A much more significant climatic reversal began around 13 000 Cal. yr BP, as temperatures declined and arctic conditions returned to the British Isles. Over Scotland, which may have been completely deglaciated during the interstadial, snow and ice accumulated once again, and by *c*.12 700 Cal. yr BP formed an ice cap several hundred metres thick over the western Highlands. This period, named the Loch Lomond (or

Table 3.4 The Late-glacial stratigraphy of northwest Europe

Age (Cal. yr BP)	Britain	Northern Europe	Climate
After 11 500	Holocene	Holocene	warm
11 500–13 000	Loch Lomond stadial	Younger Dryas stadial	cold, glacial re-advance
13 000–15 000	Windermere interstadial	Allerød interstadial Older Dryas Bølling interstadial	moderately warm a brief cool interval warm
15 000–18 000	Devensian glaciation	Oldest Dryas	cold
Before 18 000	Devensian glaciation	Weichselian glaciation	glacial

Plate 3.3 6 m core from Windermere, England laid out in 1 m-long sections, with the present-day at the top right. The pink-coloured minerogenic sediment in the nearest section was deposited immediately after the ice retreated (c.16 000 Cal. yr BP); the darker, organic muds that follow this belong to a warmer interstadial (15 000–13 000 Cal. yr BP), but the pink minerogenic sediments in the middle of the third section mark a return to a cold climate in the Younger Dryas stadial (13 000–11 500 Cal. yr BP); the last three sections of dark, organic mud cover the Holocene proper

Younger Dryas) stadial, produced some of the best-preserved glacial and pro-glacial landforms to be found in Britain. They include not only moraine ridges and eroded cirques, but also the famous 'parallel roads' of Glen Roy. The latter marked the shorelines of a former ice-dammed lake whose water levels were controlled by the height of its point of outflow. Smaller valley or cirque glaciers formed elsewhere in upland Britain, while at lower elevations periglacial processes were active. Permafrost is indicated from fossil pingos and ice-wedge polygons, and these ground conditions rapidly destroyed the partially developed interstadial soils. Lake sedimentation once more became dominantly minerogenic (see plate 3.3).

The return of a glacial climate not surprisingly threw vegetation successions into reverse. Birch woodland became restricted to locally favourable habitats, to be replaced by open-habitat taxa able to survive on disturbed and seasonally frozen soils, such as chenopods (goosefoot family). Vegetation patterns became spatially differentiated on the basis of aspect, altitude, climatic gradients and other factors (Pennington, 1977). For instance, the proportion of *Artemisia* (wormwood/mugwort) pollen at Scottish sites during the Loch Lomond stadial has been shown to co-vary with the height of the reconstructed firn line (see figure 3.4). This has plausibly been explained in terms of a marked precipitation gradient across the country, with *Artemisia* pollen most abundant in the dry, snow-shadow zone to the north and east (Birks and Mathewes, 1978; Sissons, 1980). Beetle faunas confirm pollen and geomorphic evidence that the British Isles was a tundra-like landscape in the Loch Lomond stadial, and that this was a product of a cold and generally dry climate.

Terminal Pleistocene climatic oscillation: global or regional?

It is natural that having been recognized in one area, namely northwest Europe, changes such as the Late-glacial climatic oscillation will be sought for and found elsewhere in the world. There is danger of a reinforcement syndrome here, so it is important that evidence from other areas needs to be considered carefully and on its own merits rather than on the goodness of fit with the northwest European framework.

Deep-sea sediment cores from the Atlantic show that the polar front shifted as far north as Iceland during the warmer interstadial (15 000–13 000 Cal. yr BP), only to descend southwards again to Iberia in the subsequent stadial (13 000–11 500 Cal. yr BP) (Ruddiman and McIntyre, 1981). A similar sequence is indicated on the contiguous land areas of western Europe (e.g. Turner and Hannon, 1988; Allen et al., 1996; Lowe et al., 1994) and the Alps (Lotter et al., 1992), while an increasing number of pollen diagrams from the Mediterranean basin are indicating a reversal in the readvance of woodland vegetation at the end of the Pleistocene (Rossignol-Strick, 1995; Watts et al., 1996). Further eastwards and southwards, however, the climatic signal is sometimes less pronounced than in northwest Europe (Watts, 1980; Coope and Lemdahl, 1995). Within Europe the thermal oscillation therefore appears to have varied regionally in its intensity.

In eastern North America, attempts to find an equivalent to the European Late-glacial oscillation have largely been inconclusive; the exception being along the Atlantic seaboard of Canada and New England (Watts, 1983; Mott et al., 1986; Mayle and Cwyner, 1995; LaSalle and Shilts, 1993). Lake-level and pollen data imply that cool dry events punctuated the longer-term glacial to interglacial climatic transition in the American West (Benson et al., 1997). Pollen diagrams and glacier fluctuations suggest that a terminal Pleistocene climatic oscillation also affected the cool temperate extremities of both North and South America (Peteet, 1993; Heusser and Streeter, 1980; Heusser, 1993; Clapperton, 1993). However, Coleopteran evidence for temperature reversal in the southern Andes is less clear than in Europe (Hoganson and Ashworth, 1992), implying a weaker climatic signal.

In parts of the tropics the period 15 000–11 500 Cal. yr BP was one of marked climatic instability (Gasse et al., 1990; Kudrass et al., 1991; Islebe et al., 1995). High-altitude pollen diagrams from the northern Andes show an irregular transition from Páramo grassland to Andean forest (see figure 3.3), with either one or two terminal Pleistocene stadials depending upon elevation (van der Hammen, 1974). In this case it is likely that vegetation changes were related primarily to fluctuations in temperature. However, moisture variations may have been as important as temperatures in controlling the water balance of tropical lakes. In Africa, lake-level fluctuations represent some of the strongest evidence for terminal Pleistocene climatic instability. This was a transitional period in the lake record, many lake levels rising after 15 000 Cal. yr BP only to fall again at around 13 000 Cal. yr BP (Street-Perrott and Roberts, 1983; Roberts et al., 1993). Low water levels in the tropics reflect a brief arid phase, which would appear to have been a product of the same disturbance to the climatic system that produced the northwest European Younger Dryas stadial (13 000–11 500 Cal. yr BP). There is more than a suggestion that the hydrological cycle was weakened at this time, perhaps because glacial meltwater reduced evaporation from oceanic surface waters in the northern hemisphere.

One of the most important pieces of evidence for the Younger Dryas having global consequences comes from the sea-level record. Richard Fairbanks (1989) carried out detailed dating of cores taken through drowned coral reefs in Barbados, which normally grow within a few metres of the ocean surface. From this he built up a history of sea-level rise since the last glacial maximum, which showed a significant slow-down between 13 000 and 11 500 Cal. yr BP (see figure 3.5). In this region, sea levels are primarily controlled by the eustatic factors and therefore closely reflect the build-up and decay of ice sheets. The reduced melting of the major ice sheets during the Younger Dryas, implied by the slow-down in the rate of sea-level rise, is also evident in temperature reversals in the ice-core records

Figure 3.4 The Loch Lomond (Younger Dryas) stadial in Scotland (data from Sissons, 1980 and Tipping, 1985). Coastlines are modern

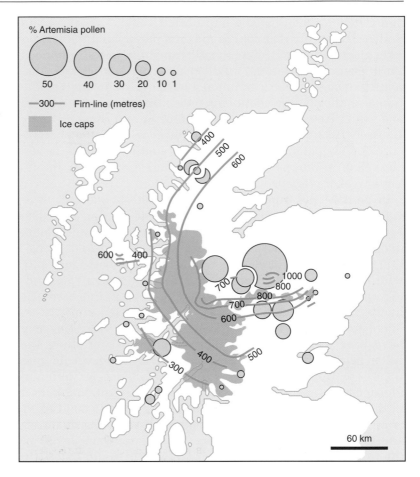

from Greenland and – although much attenuated – also in Antarctica (Mayewski et al., 1996). The Greenland cores confirm the fossil insect evidence in showing that some of the climatic changes during the last glacial to interglacial transition were startlingly rapid, including a warming of 7°C in only 50 years at the end of the Younger Dryas (figure 3.5).

Ice-core records show that temperatures changed in close harmony with atmospheric greenhouse-gas concentrations, providing evidence for a link between the two. The same correlation is evident in stomatal densities in plant leaves dating from the Younger Dryas, which are believed to be the result of atmospheric CO_2 levels (Beerling et al., 1995). Greenhouse gases are unlikely to have been the main cause of the rapid 'flip-flops' in climate during the last glacial to interglacial transition, however. Instead, it is now thought that sudden jumps in climate may be linked to periodic switches on and off of part of the 'conveyor-belt' circulation of the Atlantic Ocean (Broecker et al., 1985). The Atlantic conveyor carries warm surface water

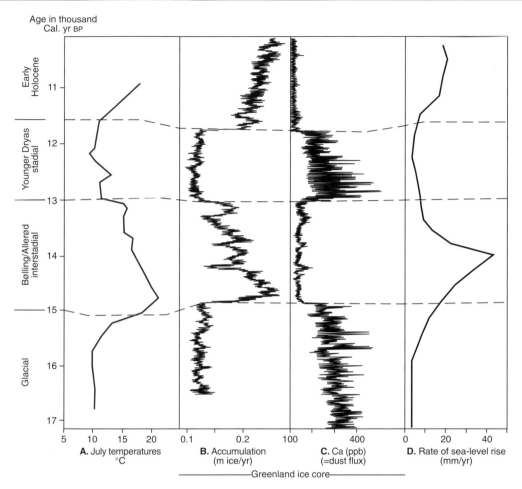

Figure 3.5   The Late-glacial climatic oscillation: **A** mean July temperatures for Britain based on Coleoptera (data from Atkinson et al., 1987; Lowe et al., 1995); **B** ice accumulation rate, and **C** atmospheric dust, in the GISP2 summit ice core from Greenland (after Alley et al., 1993; Mayewski et al., 1994); **D** deglaciation as measured by the rate of sea-level rise in Barbados corals (data from Fairbanks, 1989; Bard et al., 1990)

northwards and cold, dense deepwater southwards, keeping Europe several °C warmer than equivalent latitudes elsewhere. The present operation of the conveyor is delicately balanced depending on the salt budget of the Atlantic Ocean. A small decrease in the salinity of North Atlantic surface waters would be enough to prevent it sinking to form deepwater, the conveyor would then stop, and Europe would suddenly become much colder. Because this thermohaline circulation feeds into the rest of the world's ocean circulation, switching it on or off would also affect climates well beyond the North Atlantic (Street-Perrott and Perrott, 1990). Wallace Broecker and colleagues (1990) have proposed that fluctuations in the salt budget of the Atlantic occur naturally, leading to the 'conveyor-belt' circulation being switched off every two or three thousand years. Alternatively, large pulses of freshwater from melting North American and European ice sheets would have had a comparable effect (Broecker, 1994). The Younger Dryas climatic reversal, which bucks the Milankovitch trend, was most probably the result of a shut-down or weakening of the Atlantic conveyor coinciding with longer-term northern hemisphere deglaciation.

If the evidence for a terminal Pleistocene climatic oscillation is world wide in its distribution, then why were some areas affected whilst others were not? Probably a major factor is environmental sensitivity. For example, continental interiors and areas adjacent to large ice sheets would have been buffered against short-term climatic changes in comparison with maritime regions such as the North Atlantic seaboard. It may be significant, for example, that the Younger Dryas oscillation is clearly evident in the Summit ice cores from Greenland, but is a much more feeble feature of the Vostok cores taken in the much larger, more continental Antarctic ice sheet. Other ecologically insensitive areas would have been those far removed from plant and animal refuges. By contrast, areas near climatic limits, for instance of tree growth, would have been sensitive to relatively minor temperature and moisture variations. In combination, these factors may go some way to explaining the uneven spatial distribution of the last glaciation's final gasp.

Adjustment of geomorphic systems

Because soil formation and vegetation lagged behind the often rapid shifts in climate, many landscapes experienced a phase of temporary geomorphic instability during the terminal Pleistocene. In addition to the many landforms created directly by deglaciation, such as eskers and kettle-holes, there were also those produced indirectly, for example by glacial meltwater. For instance, meltwater flooding down the Mississippi, which reached a peak around 13 600 Cal. yr BP, reduced surface sea-water salinities in the Gulf of Mexico by 10 per cent (Emiliani, 1980). Soon after this, ice retreat opened a spillway eastwards, and meltwater instead poured through what is today the St Lawrence into the North Atlantic (Teller, 1990). This diversion of meltwater from the Mississippi to the St Lawrence may even have provided a trigger for a switching off of the Atlantic Ocean 'conveyor' during the Younger Dryas. Another, even more spectacular, pro-glacial drainage system existed in the Columbia river system in the northwest United

States, where repeated draining of pro-glacial Lake Missoula caused short-lived floods with discharges up to 20 times total modern global runoff – the largest terrestrial water flows so far known on Earth (Baker and Bunker, 1985; O'Connor and Baker, 1992; Smith, 1993).

Braided stream deposits testify that increased discharges of both water and sediment were widespread at the end of the last glaciation (Rose et al., 1980; Baker, 1983; Maizels and Aitken, 1991). These palaeohydrological changes often led to dramatic alterations in river morphology, a process termed metamorphosis by Stanley Schumm (1969). In alluvial plains, such as those of the Murray Basin in Australia, large channels containing sandy sediments were replaced by sinuous silty-clay ones at the start of the Holocene (Bowler and Wasson, 1984). Many streams and rivers which experienced increased discharges during the terminal Pleistocene are now underfit. George Dury (1964) calculated that ratios between modern and former meander wavelengths in North America vary from 1 : 10 near the ice front in Wisconsin to 1 : 3 in the more southerly Ozark Mountains. These, he proposed, reflect late Wisconsinan discharges up to one hundred times greater than at present (Dury, 1965). At a smaller scale, the dry valleys of southern England are also relict features from the Pleistocene, in this case having been produced by solifluction and related pro-cesses during seasonal snow-melt (Kerney et al., 1963).

The unvegetated glacial outwash plains of the terminal Pleistocene were also vulnerable to deflation by the wind, and aeolian sand and loess were actively transported and widely deposited as the ice sheets retreated (Catt, 1978; Wells, 1983). The small river valleys of the Dutch–Belgian border, for instance, became choked with wind-blown sand at this stage (Vandenberghe and Bohncke, 1985). After 15 000 Cal. yr BP the cold, dry climate and near-absence of vegetation gave way to the warmer conditions of the Late-glacial interstadial and river palaeohydrology changed strikingly. In the case of the River Mark valley shown in figure 3.6, sediment load diminished as aeolian activity ceased and slopes became stabilized by soil and vegetation, while at the same time river discharges increased. In combination these resulted in sharp incision of the pre-existing valley fill to depths of over 8 m. Although these incised meanders soon began to fill up with sediment, the fluvial system remained in a disturbed state with localized blocking of the valley by sand, ponding back of rivers and peat formation. Only around 10 000–9000 Cal. yr BP did the geomorphic disequilibrium of the Pleistocene–Holocene transition give way to a regular longitudinal river profile

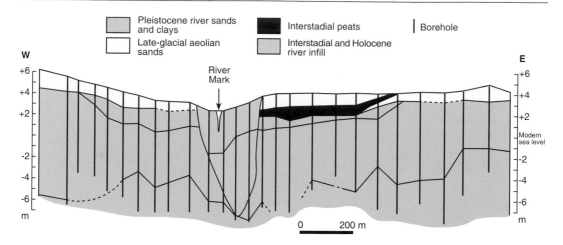

Figure 3.6 Cross-section through an alluvial valley fill, River Mark, Belgium/Netherlands (after Vandenberghe and Bohncke, 1985)

(Vandenberghe et al., 1984). In general, Holocene fluvial systems in temperate latitudes have been characterized by lower-energy environments than had existed previously, and by a sediment supply that was regulated by the well-developed forest and grassland cover (Knox, 1984). In consequence, Holocene alluvial fills in temperate latitudes are typically fine-grained in nature (Gregory et al., 1987).

Intense geomorphic activity during the Pleistocene–Holocene transition also characterized many low-latitude regions. Although the climatic transition here was from arid to humid rather than from periglacial to temperate, the same kind of disequilibrium between vegetation and climate was created. Unvegetated, erodible soils were combined with high-rainfall erosivities to produce the highest sediment yields of the whole late Quaternary prior to recent human impact. These yields are recorded by increased rates of sedimentation in coastal deltas, inland lakes and river terrace sequences (Roberts and Barker, 1993; Thomas and Thorp, 1995). The Nile, which had been a highly seasonal, braided river between 23 000 and 15 000 Cal. yr BP, experienced a huge increase in discharge, partly on account of overflowing East African lakes expanding the river's catchment area.

Human Ecology at the End of the Pleistocene

The probable birthplace of the human race lies in East Africa, around 5 million years ago. From here the genus *Homo* spread into all other continents save Antarctica by the start of the Holocene. However, establishing the precise date at which hominid populations first arrived at a particular part of the Earth's surface is problematic, not least because new discoveries are constantly pushing these dates further back in time. It

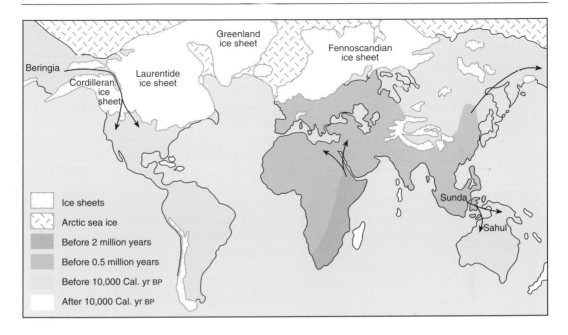

Figure 3.7 The human colonization of ice-age Earth

is perhaps safer to consider the latest agreed rather than the earliest possible dates of arrival, and using this approach a three-stage model for human expansion may be suggested. For the first few million years, hominids were restricted to eastern and southern parts of the African continent, but by no later than half a million years ago they had spread into the adjacent continents of southern Asia and Europe.

The second stage in the expansion came significantly later and involved the peopling of Australia and the Americas. In both cases there were formidable natural barriers to be overcome which delayed human colonization until late in the Pleistocene. To reach Australia–New Guinea meant a sea crossing by raft or canoe of at least 100 km even at times of low sea level. Remarkably, this was achieved at least 40 000, and possibly more than 100 000 years ago, as indicated by ^{14}C- and luminescence-dated burial sites such as Lake Mungo (Thorne, 1980; Jones, 1989; Fullagar et al., 1996). By 30 000 years ago all of Australia's main ecological zones had been occupied, along with the adjacent islands. The palaeo-Indians arrived somewhat later in America than the Aboriginals did in Australia. It was possible to walk from Asia to America via the dry-land bridge of Beringia (Elias et al., 1996), but once in Alaska, any potential colonizers would have found their way blocked by the giant North American ice sheets (see figure 3.7). Nor would further progress have been encouraged by the presence of some ferocious ice-age predators roaming the Yukon and Alaskan tundra, including a gigantic carniverous bear (*Arctodus simus*) (Kurtén and Anderson, 1984; Geist, 1989).

By the time the route south opened up, rising sea levels had
cut the way off back to Asia. It is therefore not entirely surprising that the main expansion of human population in the
Americas did not 'take place until the terminal Pleistocene
14 000–13 000 years ago. On the other hand, there is an active
debate as to whether the hunters of this Clovis culture were
home grown from pre-existing, pioneering groups, for which
there is some fragmentary archaeological evidence, or whether
they were of exogenous origin, descending from Alaska via an
ice-free corridor that opened up during deglaciation (West,
1983).

The third and final stage in the colonization of the Earth
took place in the Holocene and involved the inhospitable terrain around the ice sheets of Greenland and northern Canada,
and many of the world's islands. It is curious that despite the
existence of water-borne craft, virtually no Mediterranean islands were settled until early in the Holocene (Cherry, 1981;
Schule, 1993). More easily understood is the late arrival of
humans in the geographically isolated islands of the eastern
Pacific (*c.*3000 Cal. yr BP), Madagascar (*c.*2000 Cal. yr BP) and
New Zealand (*c.*1000 Cal. yr BP).

The Pleistocene colonization of the Earth by humans was an
extraordinary achievement (Cavalli-Sforza and Cavalli-Sforza,
1995). What is more, it was achieved with subsistence economies entirely based on hunting of wild animals, fishing, and
gathering of wild plants. Hunter-fisher-gatherer (h-f-g) peoples
need to have an intimate knowledge of their natural environment upon which they depend for food and other resources.
'Man the hunter' must be familiar with the movement and
behaviour of his quarry, whether stag or seal, just as 'woman
the gatherer' must know of the location and uses of different
plants, whether hemlock or hazelnuts. Ethnographic studies of
modern h-f-g groups suggest that plant foods are normally a
more important component of the diet than meat, although
both this and the gender-based division of labour may not
have been true of all Pleistocene populations. Establishing the
relative importance of the hunted/scavenged versus the gath-
ered food component at archaeological sites is not easy, because plant remains are much less well preserved than animal
bones. However, it is likely that meat provided the bulk of the
food supply to later Pleistocene populations in mid-latitude
Eurasia (Klein, 1980). This is because the open steppe-tundra
glacial vegetation would have supported large numbers of game,
but provided little in the way of edible plants. Hunting groups
were able to follow animal herds across the open landscape,
relying on them to provide protein, fat, hide, bone, and all the
other material necessities for life. By contrast, the forest environments of the tropics and of the Holocene possessed a wide

range of potential food plants which may have shifted the dietary emphasis away from meat.

H-f-g bands move camp throughout the year following game, water or other resources, and need a relatively large territory to support them, at least 75 km^2 for a band of only 25 persons. The resulting low population densities mean that even at the end of the Pleistocene, world population is unlikely to have risen above 10 million. Because this level was maintained for many thousands of years, however, 90 per cent of all humans ever to have been alive were hunters, fishers or gatherers.

Simple stone-age technology and low population numbers limited the extent of human impact upon the natural environment. Plant resources were manipulated by agencies such as fire, and burning to clear vegetation or encourage regrowth is certainly one of our oldest tools of environmental management (Lewis, 1972). Although forest or grassland fires only take hold where there is abundant dry plant matter, the spark to light them could be caused by lightning strikes as well as by deliberate human action. It is therefore difficult to know how far former fire frequencies were naturally, and how far culturally, determined. Interestingly, in a long pollen record from northern Australia, charcoal frequencies are many times higher during the Holocene than during the last interglacial, when Australia's aboriginal population had probably not yet arrived (Kershaw, 1986). However, more clearly evident than the effect on vegetation was that on other parts of ecosystems, notably upon large mammals who were the prey or competitors of *Homo sapiens*.

Megafaunal extinctions

Towards the end of the Pleistocene there occurred a devastating wave of animal extinctions. The most obvious victims were mammals with body weights over 44 kg, such as the mastodont (*Mammut americanum*), the woolly rhino (*Coleodonta antiquitatis*), the giant deer (*Megaloceros*) and the native American horse (*Equus occidentalis*) (see plate 3.4), but birds and smaller mammals were not entirely unaffected either (Grayson, 1977). The extinctions varied in their timing and severity across the globe, but the most widespread phase took place during the terminal Pleistocene (16 000–11 500 Cal. yr BP) and affected the Americas and northern Eurasia (see table 3.5). This was of course a period of rapid climatic change, and it has been suggested that many large mammals were unable to adapt to the new environmental conditions (Reed, 1970). For example, a family of mammoths (*Mammuthus primigenius*) whose remains have been found at Condover in England have been ^{14}C dated to precisely the period of most rapid warming just after 15 000

Plate 3.4 Skeleton of a woolly mammoth, one of many species of megafauna made extinct during the last glacial-to-interglacial climatic transition. This specimen, found on the banks of the River Shandra in eastern Siberia, is the largest complete skeleton so far found (the distance between the tusk ends is 3 m!). In addition to the skeleton, this mammoth's heart, liver, stomach and intestines were also preserved as a frozen mass weighing 700 kg

Cal. yr BP (Coope and Lister, 1987). Herbivores such as these found their tundra-steppe environment being displaced northwards and reduced in extent by the invasion of woodland. In turn, carnivores like the sabre-tooth tiger (*Smilodon*), who preyed upon grazing mammals, found their food resources diminishing. In some cases, movement to new areas was hampered by mountain ranges, rising sea levels or other natural barriers. This combination of climatically caused environmental stresses could, it has been suggested, have led to the more specialized and less adaptable species being exterminated.

Unfortunately, this hypothesis has some manifold weaknesses. Most obvious of these is the fact that no comparable extinctions took place at other times of rapid climatic change during the Pleistocene; for instance, at the beginning of the last interglacial *c.*130 000 years ago. Researchers such as Paul Martin have instead been struck by the coincidence between cultural changes and the timing of megafaunal extinctions. This is most clearly the case in North America, where the appearance of the Clovis culture at 13 500 Cal. yr BP was followed within less than a thousand years by the demise of many large mammals. It has been argued that big game hunt-

Continent	Extinct	Living[a]	Total	Extinct (%)	Major period of extinction (Cal. yr BP)
Australia	13	3	16	81	30–17 000
South America	46	12	58	80	15–9000
North America	33	12	45	73	16–11 500
Europe[b]	9	14	23	39	16–10 000
Africa	7	42	49	14	14–10 500

Table 3.5 Late Pleistocene extinct and living genera of large terrestrial mammals (>44 kg adult body weight)

[a] or extinct in historical times
[b] excluding Mediterranean islands
Source: Martin and Klein (1984)

ers of the terminal Pleistocene butchered, in vast numbers, animals that were quite unused to human predation (Martin and Klein, 1984). Archaeological support for this overkill hypothesis comes from 'kill' sites at which a great many animal bones are present, although perversely such associations are much more common in the Old World than the New. At Solutré in France, for instance, the bones of over 100 000 horses (*Equus przewalskii*) have been found at the base of a cliff over which they appear to have been driven by Late-glacial hunters. Later Holocene island extinctions, such as that of the giant flightless moa bird (*Dinornis*) (plate 6.3), similarly followed within a few centuries of the arrival of humans, in this case the Maori peoples of New Zealand (Cassels, 1984). Evolution in isolation meant that prior to the Holocene the Mediterranean islands were home to creatures almost as bizarre as those found in Dr Doolittle's zoo, including flightless swans, dwarf 'antelopes' (*Myotragus*) and pygmy hippos. With the arrival of the first permanent human populations at the beginning of the Holocene, all of these rapidly became extinct (Lewthwaite, 1989; Simmons, 1991; Schule, 1993). Not a single one of the 'wild' animals on islands such as Mallorca and Cyprus is therefore truly native; all have been introduced by people.

Although there are good grounds for believing that human agency was involved in most cases of faunal extinction, overkill seems no more satisfactory a universal explanation than does climatic change. Mammalian extinctions may have followed soon after the appearance of the Clovis hunters in America, but in Europe and Australia hunting populations coexisted with megafauna for many thousands of years before the latter declined at the end of the Pleistocene. At Lancefield Swamp in Australia, a bone bed contains the remains of some 10 000 giant extinct animals (Gillespie et al., 1978). ^{14}C dating indicates that humans and fauna lived together in Australia for at least 8000 years, and therefore rules out any possibility of rapid, first-contact overkill in this continent. With animals such as the giant deer (*Megaloceros*), which was abundant in Ireland during the Late-glacial period, extinction seems to have occurred before humans had even arrived.

Another feature inconsistent with overkill is that many of the animals which are most numerous at archaeological sites in fact avoided extinction, for instance the bison (*Bison bison*) in North America and the reindeer (*Rangifer tarandus*) in Europe. These lines of evidence suggest that indirect human action may have been as important as direct kill-off (Krantz, 1970). As hunters, late Pleistocene humans would have partly replaced the ecological niche of carnivores, and may also have threatened non-favoured herbivores such as the American Shasta ground sloth (*Nothrotheriops shastense*). In contrast, favoured game species would have been preserved, probably by laws and taboos similar to those recorded historically for hunter-gatherer groups. This would have applied especially to herd ungulates like the bison and the reindeer, with whom human populations were able to develop a symbiotic relationship (see figure 3.8).

Having modified the structure and stability of ecosystems, human populations would have made other large mammals more vulnerable to environmental stress. In particular, the changing and, from the faunal point of view, deteriorating climate towards the end of the last glaciation would have forced animals to compete with each other and with humans for increasingly scarce resources. In Australia, climatic desiccation from about 30 000 to 18 000 Cal. yr BP forced aboriginal and animal populations into competition for food and water resources, with the latter suffering in consequence. In western North America, a similar contraction of water resources occurred at around the time of the Younger Dryas, with Clovis culture artifacts and mammoth tracks being found together around spring sites (Haynes, 1991). Indirect human disturbance of ecological balances is consistent with the loss of birds and small mammals, and also with the differing severities of late Pleistocene extinctions across the world. In Africa, where hominids had formed part of savanna ecosystems for millions of years, a phase of extinctions can hardly be recognized at all. Nearly all of that continent's megafauna has remained intact until the present day, to fill the game parks and cages of modern zoos. In northern Eurasia some eight genera disappeared at the end of the Pleistocene, all of them adapted to steppe-tundra environments and most exploited by stone-age hunters through the last glaciation. In the Americas, where human antiquity is shorter and ecosystems were almost pristine until the terminal Pleistocene, mammalian extinctions were devastating in their effects. They involved 39 genera of large mam-

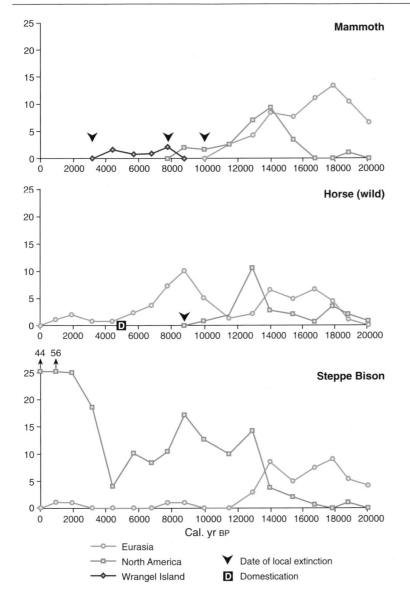

Figure 3.8 The number of dated finds of three megafauna in the New and Old Worlds (data from McDonald, 1984; Dolukhanov and Khotinsky, 1984; Vereshschagin and Baryshnikov, 1984; Vartanyan et al., 1993)

mals in North America alone, a greater number of extinctions than during the preceding 4 million years (Martin and Klein, 1984), and have left New World megafaunas impoverished ever since.

Perhaps the most remarkable discovery of all has been the fossil remains of dwarf mammoths of Holocene age on Wrangel Island in the Siberian Arctic (Vartanyan et al., 1993). During the late Pleistocene, Wrangel Island was joined to the mainland by the land bridge of Beringia, but, as sea levels rose, so it became isolated, and its mammoth population was trapped. Elsewhere, the mammoths quickly became extinct, but here they were able to survive in an ice-age time capsule of tundra meadows and arctic steppe, untroubled by natural predators or by humans. As often happens with isolated populations, the mammoths started to 'shrink' in size, possibly to around half their previous height. Numerous bones, tusks and teeth of these dwarf mammoths have been ^{14}C dated to the early and mid-Holocene, but – sadly – they show that the mammoth population eventually succumbed, almost certainly to natural causes. The youngest dates, however, indicate that on this remote Arctic island, mammoths survived until after 4500 years ago, by which time Egypt's great pyramids were being erected!

The human role in terminal Pleistocene extinctions remains a matter of controversy. None the less, it is likely that cultural impact upon at least this part of ecosystems was substantial even before the start of the Holocene. The separation of human and natural agencies therefore emerges as a problem even before climate, vegetation and landscapes had become recognizably modern in form.

CHAPTER FOUR

Early Holocene Adaptations (11 500– 5000 Cal. yr BP)

Changes in the Physical Environment

The onset of the Holocene witnessed the start of environmental processes which have continued up to the present day; processes such as soil formation, plant succession, lake ontogeny and faunal migration. Although interglacial conditions became established soon after 11 500 Cal. yr BP, it would be quite wrong to imagine that the climates, ecosystems and landforms of the early Holocene were identical to those of the present day or that they have remained static since that time. The broad outlines of the modern natural world were established earlier in some regions and later in others, and for a time this produced strong inter-regional contrasts, sometimes even over short distances. The first set of changes to be considered will be those taking place in the physical environment during this formative stage of the Holocene.

Ice sheets and sea levels

Over northwest Europe and many other parts of the world, modern summer temperatures were established within the first 1000 years of the Holocene (see figure 3.5, p. 75), and this temperature rise produced a rapid response in the smaller ice caps and valley glaciers. In the Val d'Hèrens in the Swiss Alps, for example, a moraine dated to 9400 Cal. yr BP lies adjacent to the modern Tsidjiore Nouve glacier, marking a retreat of over 20 km from the position of the glacier snout two millennia earlier (Rothlisberger and Schneebeli, 1979) (see plate 2.6, p. 48). However, because of their very bulk, the major northern hemisphere ice sheets took longer to melt away. In particular, the Laurentide ice sheet remained extensive until after 9000 Cal. yr BP, when its heart was eaten out as the sea invaded Hudson Bay.

The retreat of the ice sheets caused eustatic sea levels to rise from about −55 m at 11 500 Cal. yr BP to around modern elevations c.6000 years ago (Tooley and Shennan, 1987; Fairbanks, 1989; Chappell and Polach, 1991). Up until the end of the Pleistocene, eustatic sea-level rise had accounted for most coastline changes because of the virtually instantaneous response of eustasy to ice-sheet melting. Glacio-**isostatic** adjustment in deglaciated regions involved a slower response, but it has played a prominent part in the Holocene history of European and North American coastal zones (see figure 4.1). In places previously compressed under ice, the unloading of this weight caused the land literally to rebound out of the sea (see plate 4.1). The scale of changes since the terminal Pleistocene can be judged from the fact that Late-glacial shorelines in Scotland

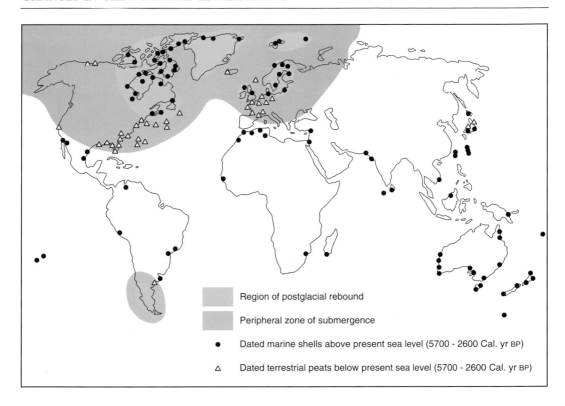

Region of postglacial rebound

Peripheral zone of submergence

● Dated marine shells above present sea level (5700 - 2600 Cal. yr BP)

△ Dated terrestrial peats below present sea level (5700 - 2600 Cal. yr BP)

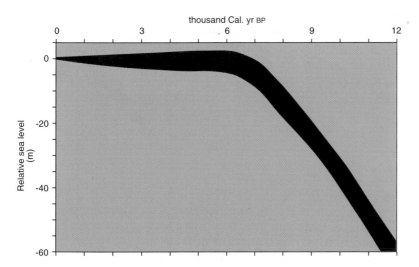

Figure 4.1 Holocene sea-level changes: **A** areas of rebound and submergence (after Walcott, 1972); **B** eustatic sea-level curve (data from Bard et al., 1996)

are now up to 40 m above sea level, representing a glacio-isostatic uplift of a least 70 m during the past 15 000 years, while in some parts of Scandinavia, Holocene uplift has been many times that figure.

Adjacent to the area of former glaciation, on the other hand, was a zone whose surface had actually been pushed up by the weight of the ice. This forebulge has experienced compensatory subsidence during the Holocene and forms a zone which continues to suffer submergence today. It is the increasing risk of inundation along the US eastern seaboard and the southern North Sea littoral that has led to the construction of major sea defence systems such as have been built in the Rhine delta region of the Netherlands.

The interaction of eustatic and glacio-isostatic factors has often led to complex local histories of relative sea-level change, none more so than in the case of the Baltic. A large pro-glacial lake formed here at the end of the Pleistocene, adjacent to the retreating Fennoscandian ice mass. Towards the tail end of the last glaciation, the ice retreated from Mt Billingen in south central Sweden and caused the Baltic ice lake to fall to what was then sea level (Björck and Digerfeldt, 1989). The Baltic became a marine embayment, named the Yoldia Sea after a brackish-water mollusc that lived in its waters. Glacio-isostatic uplift, however, outpaced the eustatic sea-level rise in this region and isolated the Baltic once more. This second fresh-water body formed *c.*10 600 Cal. yr BP and has been named the Ancylus lake (Eronen, 1983). Finally, *c.*8400 Cal. yr BP, the sea once more succeeded in flooding what are now the Straits of Denmark to produce the Littorina Sea (named after the common periwinkle), the immediate predecessor of the modern Baltic.

The configuration of the glacial coastline differed most strongly from that of today in areas with shallow offshore gradients. It was these same areas which were now most rapidly inundated by the incoming tide. The subcontinent of Sunda in southeast Asia, for example, became fragmented into the many islands of the Indonesian archipelago. In areas such as these, land was drowned at a rate that must have been noticeable from one year to the next. What is sure is that human populations had to relocate themselves and their economic activities, as archaeological sites adapted to the life of the seashore, such as shell middens, testify.

Human adaptations to coastal environments

Human utilization of coastal resources certainly extends well back into the Pleistocene. On the other hand, most evidence from these early periods has been drowned by Holocene high sea levels, so that it is only for the late Quaternary that we have an adequate record of coastal occupation patterns. In Japan, the Jomon Neolithic culture, which endured from 11 500 until as late as 2350 Cal. yr BP, represents one such adaptation to the diverse marine and littoral resources of the Holocene (Glover, 1980). Around 8000–7000 years ago the Jomon marine transgression, of tectonic as well as eustatic origin, produced sea levels 3 to 6 m above modern elevations (Fujii and Fuji, 1967). Along this former coastline, now up to 60 km inland, lie shell mounds and other Jomon occupation sites containing marine molluscs such as *Meretrix* and *Mactra*, and remains of deep-sea fish (Akazawa, 1986). Rising Holocene sea levels also served to isolate the Japanese islands from the rest of East Asia. The Jomon Neolithic, with its distinctive material culture that included pottery, is certainly culturally different from any of the contemporary developments on the Asian mainland.

In other cases the physical isolation brought about by rising sea levels led to cultural and economic retardation. Tasmania is a case in point. This continental island was joined to the Australian mainland during glacial low sea levels, but after 14 000 Cal. yr BP this Bassian land bridge was severed (Jones, 1977). Although the indigenous Tasman aboriginals possessed water craft made of bark fibre and utilized coastal resources such as shellfish, they were unable to cross the stormy seas of the Bass Strait, today 250 km wide. The Tasman population did adapt to insular life and survived until their extinction following upon European contact, but ethno-archaeological work has revealed that during the Holocene they 'lost' the use of many attributes including the boomerang, the concept of hafting, and the habit of eating fish (Jones, 1977, p. 343). Nor did any of the technological innovations which occurred on the mainland after 14 000 Cal. yr BP reach Tasmania. It seems as if the rising post-glacial sea served to isolate Tasmanian society completely from mainstream aboriginal developments, producing an idiosyncratic culture analogous to the unique bird and animal adaptations found on some oceanic islands.

One of the best-documented adaptive responses to rising sea levels comes from Franchthi cave in southern Greece. This site, excavated under the direction of Tom Jacobsen since 1967, has an unusually long, unbroken archaeological sequence beginning before 15 000 and ending around 5800 Cal. yr BP. Today the cave entrance lies only a few metres above sea level on a rocky coast that can be reached only by boat. But as studies by Tjeerd van Andel have shown, the situation was formerly different. At the start of the Holocene the sea lay 2–3 km away and most of the shore comprised a flat alluvial coastal plain. Seafood products, which had been largely absent from Pleistocene deposits in the cave, began to appear

Plate 4.1 Raised marine platform, western Islay, Scotland, now 35 m above sea level after glacio-isostatic uplift

around this time, and during the Mesolithic occupation assumed a major role in Franchthi's site economy (see figure 4.2). Initially the inhabitants were mainly content to collect shells, notably of *Cyclope neritea*, a small mollusc found on muddy nearshore and estuarine flats (Shackleton and van Andel, 1980; Shackleton, 1988). But they soon ventured further afield. Numerous bones of large fish such as tunny, and the use of obsidian (a black volcanic glass) from the Aegean island of Melos, suggest deep-water fishing and sea-going boats. Around 9000 Cal. yr BP continuing sea-level rise appears to have drowned *Cyclope*'s mud-flat habitat, since it is abruptly replaced in Franchthi's late Mesolithic levels by another mollusc – *Cerithium vulgatum* – characteristic of rocky shores.

As well as reflecting changes in the proximity and hence the level of the sea, fish and shellfish remains from Franchthi may also offer a clue to the past productivity of Mediterranean waters. During the glacial-to-interglacial transition, the almost land-locked East Mediterranean basin received a large influx of freshwater. Initially this came as meltwater from the retreating Eurasian ice sheets which cascaded from the Siberian lowlands via the Aral, Caspian and Black Seas into the Aegean (Grosswald, 1980). Later it originated from floods of a 'high and wild' stage of the River Nile around 10 000 Cal. yr BP (Adamson et al., 1980). This influx formed a freshwater 'lid'

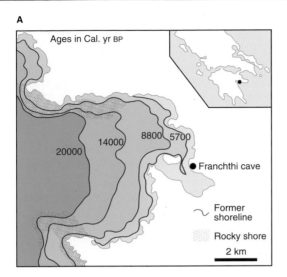

A

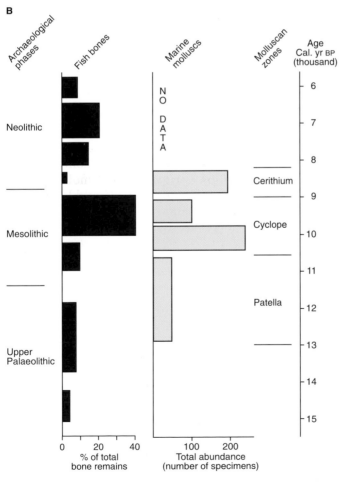

B

Figure 4.2 Franchthi cave, Greece: **A** sea-level transgression (after van Andel et al., 1980); **B** sequence of marine resource usage (data from Payne, 1975; Shackleton and van Andel, 1980)

over the main body of saline sea water which became stagnant and anaerobic, allowing organic-rich sapropel muds to be deposited on the sea bed (Williams et al., 1978; Fontugne et al., 1994). The last East Mediterranean sapropel layer is dated to 10 000–8000 Cal. yr BP, and this may have been a period when levels of nutrients in sea water and marine productivity were higher than today. It was certainly the time when marine resources were most in evidence at Franchthi cave.

Further to the north, eustatic sea-level lowering had caused the Black Sea to be isolated from the world ocean, because the Bosphorus Straits which connect them are only about 50 m deep at present. In consequence, the Black Sea became a giant freshwater lake in glacial times, and the Bosphorus was transformed into an overflow channel down which poured the excess meltwater derived from ice sheets in its catchment. By the start of the Holocene, however, the ice sheets had retreated north and the discharge of water into the Black Sea was much reduced. Like any non-outlet lake, the Black Sea responded by lowering its water level, probably falling by more than 150 m to judge by a widespread erosion surface on the coastal shelf. In the meantime, world sea levels were rising so that they became higher than those in the drawndown Black Sea. Finally, around 7500 Cal. yr BP, sea water poured into the Black Sea through the Bosphorus in a flood which may have been several hundred times greater than the world's largest waterfall today (Ryan et al., 1997). Human populations around the former Black Sea coast would have found the sea advancing landward at more than 1 km every day. The modern view across the Bosphorus at Istanbul is spectacular enough, but it is nothing compared to the show that was on offer for a few years early in the Holocene, when an estimated 50 km^3 of water per *day* cascaded in!

Lake ontogeny and soil development

Pleistocene glaciation not only affected sea levels but was also a most effective lake-producing agency, for as the ice sheets retreated innumerable lakes were left behind in North America and northern Europe. Some, such as the Laurentide Great Lakes, were large and deep, but many more were simply small water-filled hollows, such as kettle-holes. Starting life as pools of infertile water underlain by till or rock, these young ecosystems have matured during the course of the Holocene. Lake maturation or **ontogeny** has classically been viewed in terms of a series of stages, notably from an initial state of infertility or **oligotrophy**, through **eutrophy** to extinction (see Technical Box IV). It is certainly true that in small lakes, infilling by sediment progressively reduces water depth, while encroach-

Technical Box IV: Limnology

The study of lakes – or limnology – provides a good illustration of how physical, chemical and biological processes are interrelated within a single ecosystem (Hutchinson, 1957–75). Lakes may be large or small, shallow or deep. The water of lakes that are deep relative to their surface area will not be able to circulate completely for at least part of the year. In particular, only the surface water of temperate lakes is warmed up in summer to create an epilimnion zone which is separated from the cold lower waters – or hypo-limnion – by a sharp transition known as the thermocline (see figure 4.3). In very deep, meromictic lakes such as Tanganyika, the thermocline persists all year round and the top and bottom waters never mix. More usually, lake waters are separated for only part of the year, with seasonal overturn of top and bottom water. These are termed dimictic lakes. In the dark waters of the hypolimnion, photosynthesis may be negligible and the store of oxygen will become depleted, to create an anaerobic environment.

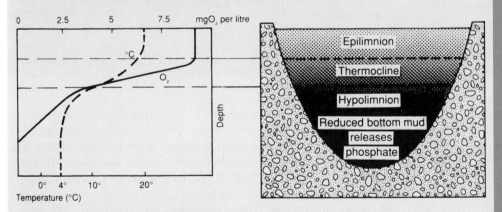

Figure 4.3 Temperature and oxygen profiles of a stratified eutrophic lake in summer

Lakes may be further characterized on the basis of their primary biological productivity. In those receiving ample nutrients from their catchments, algal productivity is relatively high and the lake eutrophic. Nutrient-poor lakes, by contrast, are much less productive and are described as oligotrophic. De-oxygenation of the hypolimnion is much more marked in eutrophic than in oligotrophic lakes, partly because of the continual 'rain' of dead organic matter derived from algal growth in the epilimnion falling towards the lake bed. The bottom sediments of eutrophic lakes therefore have a higher organic content than those of oligotrophic lakes. Under the reducing conditions typical of eutrophic bottom waters, phosphorus is released back into the lake water from the bottom sediments, maintaining the nutrient supply.

TECHNICAL BOX

Plate 4.2 Blelham Tarn, English Lake District, with a hydrosere sequence at the lake edge

ment of plants from the shore reduces lake area. This succession from open water to reed swamp to fen is termed a hydrosere, and like other ecological successions is often described in terms of a predictable orderly sequence of plant communities (see plate 4.2). Donald Walker (1970) analysed the sediment and plant macrofossil records of cores taken at the edge of infilling lakes across the British Isles, and found that they rarely show such an orderly sequence from open water to a dry-land plant community. Placing 12 possible communities in sequential order he found that most successional steps involved at least one seral stage being missed out. Not only were stages jumped, but in a significant number of cases there was a reversion from a 'later' seral stage such as fen carr to an 'earlier' one such as reed swamp. This more complex vegetation sequence during lake infilling suggested that factors external to the lake, such as variations in regional climate, were at least as important in determining the ecological pathway as were autochthonous ones within the lake.

Other evidence which initially gave support to the stage model of lake evolution came from an increase in organic matter at the base of the sediment infill in many temperate lakes. Ed Deevey (1942) interpreted this increase as reflecting a sigmoid (or S-shaped) growth in biological productivity at the start of the Holocene. However, other palaeolimnologists such as Dan Livingstone (1957) have questioned whether this organic carbon necessarily resulted from autochthonous production by algae. Organic matter can also originate from leaching or erosion of catchment soils, and may, in any case, be diluted at times of high mineral inwash, such as during the terminal Pleistocene. John Mackereth working in the English Lake District suggested that lake muds could be better thought of as 'a series of samples of soils eroded from the drainage basin and deposited chronologically in the lake bed' (1966). If so, then it may be possible to reconstruct the post-glacial development of temperate soils by analysing the chemistry of lake sediments.

Mackereth hypothesized that the importance of the inorganic component of lake sediments would reflect the intensity of erosion in the catchment. In an actively eroding landscape, unweathered minerals such as Na (sodium) and K (potassium) would be brought into the lake. This occurred in Lake Windermere during the Late-glacial stadial (13 000–11 500 Cal. yr BP) when pink minerogenic clays, low in organic matter, were deposited. On the other hand, if the catchment were stabilized by soil and vegetation, runoff would instead carry dissolved ions and organic matter into the lake. Some of these would in due course be incorporated into lake-bottom sediments, either directly or through aquatic organisms such as algae. The switch to organic sedimentation of this kind in Windermere began abruptly at the start of the Holocene (see plate 3.3, p. 71). We can therefore infer that in the protocratic phase catchment soils were of base-rich, brown earth type, with accumulation of mull humus and nitrogen.

Although many lake ecosystems did experience high productivities during this protocratic phase, they were rarely maintained subsequently. Once limiting nutrients, such as P (phosphorus), had been released and used up, and once networks of ecological competition and predation were established, most lakes reverted to a mesotrophic or oligotrophic condition (Birks, 1986a). The trend towards declining productivity and nutrient status continued during the mid-Holocene as soils became progressively leached. According to Svend Andersen's (1969) soil retrogression model, pedogenic changes from neutral to acid soils is an inevitable feature of these later stages of interglacials in many areas of moist, temperate climate (see figure 3.1, p. 59). In brown earths, litter from the mixed deciduous forest is rapidly broken down by soil organisms such as earthworms. However, with leaching of bases such as calcium, earthworms are replaced by arthropods (insects and spiders), decomposition of litter is no longer complete, and acid mor humus begins to accumulate. Fe (iron) and Al (aluminium) are dissolved and moved down-profile to leave bleached sands over an impermeable iron-pan horizon – a typical podzolic profile. In some upland areas this was the precursor to waterlogging and the formation of extensive blanket bogs (see chapter 6).

Mackereth (1966) thought soil podzolization should be recorded in lake sediments, specifically in the flux of iron and manganese (Mn). These elements are found in both mineral and organic soils, but they are relatively insoluble under oxidizing conditions. On the other hand, under reducing conditions brought about by impeded drainage or the build-up of mor humus, Fe and Mn can be released from the soil in solution. Mackereth used the abundance of these elements in lake muds to reconstruct changing soil conditions. He argued, for instance, that with Mn being more soluble than Fe, the low Fe : Mn ratio in the sediments of Lake Windermere was a reflection of reducing conditions in the soils of its catchment, while the higher Fe : Mn ratio in Ennerdale Water was due to that lake's thinner, mineral soils. Other formerly glaciated upland areas experienced similar iron enrichment associated with podzolization and leaching of soil humus, to create brown-water, dystrophic lakes. In Labrador, Canada, this occurred as boreal forest replaced shrub tundra between 8400 and 7800 Cal. yr BP (Engstrom and Wright, 1984). Palaeolimnology therefore indicates that in areas of acid bedrock, mineral soils were replaced by increasingly well-leached ones during the Holocene, as Andersen's model predicted. The accompanying lowering of soil pH values had downstream consequences for streams and lakes, many of whose waters experienced a progressive, natural acidification over this time.

In areas like the northeastern United States and central Europe, which had supported boreal woodland in Late-glacial times, soils evolved in the opposite direction at the start of the Holocene – from podzol to brown earth. In eastern Hungary, for example, Kathy Willis used a combination of pollen and sediment chemistry from the same lake core to examine how vegetation and catchment soils changed as the climate warmed (Willis et al., 1997). At the start of the Holocene this area had supported an open coniferous forest-steppe, but at about 9500 Cal. yr BP the vegetation switched within no more than a century to mixed deciduous forest (figure 4.4). Chemical analysis of the lake sediments shows that there was also release of chemical elements, such as iron and manganese, which are typically associated with podzolic soils, followed by a sustained increase in calcium and reflecting the development of more alkaline, brown earth soils in the catchment. Significantly, these two sets of changes did not occur at exactly the same time, with the soil change lagging several centuries behind the arrival of deciduous trees. While there were undoubt-

edly a number of factors involved in the replacement of podzols
by brown earth soils, such as wildfires (as reflected in a peak
in charcoal), it is clear that here the post-glacial increase in
deciduous trees was a cause, and not a consequence, of changes
in soil type.

The Holocene evolution of lowland temperate lakes was never
so strongly limited by nutrient availability as were the acid,
oligotrophic waters of the uplands. Lakes such as Lake Erie
(US/Canada) and Lough Neagh (Ireland) consequently became
mesotrophic or eutrophic. It is apparent that the ontogeny of
individual lakes depended on the particular combination of
factors – geology, vegetation composition, soil development,
lake morphometry, catchment-lake (z) ratio – that applied in
each case. Some combinations produced pathways leading to
dystrophy, others to eutrophy, yet others to infilling and extinction (Deevey, 1984).

The Return of the Forests

During the cold climate of the last glacial stage, mid-latitude
deciduous and boreal woodland were located nearer to the
equator, often – as in the case of Europe – being forced into
isolated refuges. Similarly, the tropical rainforests of Africa
and Amazonia were reduced in extent and fragmented by climatic aridity. The subsequent return of forest ecosystems at
the end of the Pleistocene and the beginning of the Holocene
is one of the great stories of the Earth's recent natural history
and it is well documented by palynology. In particular, the
many well-dated pollen diagrams from Europe and eastern
North America make it possible to map the direction and rate
of movement of both individual plant taxa and biomes across
these two sub-continental land masses (Huntley and Birks,
1983; Webb et al., 1993; Prentice et al., 1996).

Europe

Perhaps surprisingly, Europe's glacial tundra-steppe and boreal
forest remained essentially unchanged until almost 11 500
Cal. yr BP, at least when viewed on a sub-continental scale.
The restricted response of flora to Late-glacial temperature
fluctuations highlights the time lags that can occur between
climate change and plant distribution. For example, despite
experiencing temperatures close to those of today during the
Late-glacial interstadial, the British Isles remained a largely
open landscape until the start of the Holocene. By this point
in time, however, deciduous trees had begun to expand out
of their southern refugia and ground conditions had been

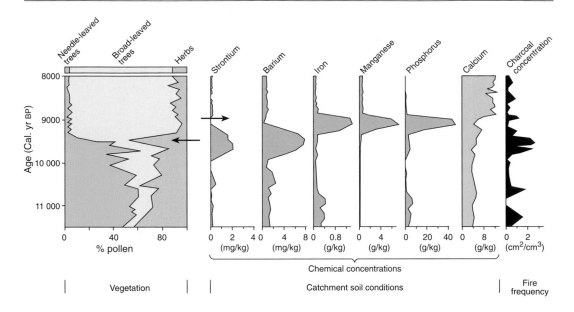

Figure 4.4 Early Holocene vegetation and soil changes reflected in a sediment core from Hungary (based on Willis et al., 1997). The arrows mark the main points of change first in vegetation, and subsequently in soils

prepared by pioneering plant colonizers. Consequently, once climatic conditions permitted, there were few checks to a rapid spread of tree species across Europe, other than their own rates of dispersal. Between 11 500 and 9000 Cal. yr BP tree species moved in to occupy suitable vacant land with much the same self-interested zeal as the miners of the '49 Gold Rush. The result was no less chaotic, with short-lived, local-ized associations of convenience formed between species; associations which often have no contemporary equivalent and hence no modern analogue. An example of the vegetation of this initial protocratic phase of the Holocene would be the sub-arctic birch-aspen woodland that existed adjacent to the retreating southern margin of the Scandinavian ice sheet. Stud-ies of absolute pollen influx show a marked change during the protocratic phase, primarily because of a huge increase in vegetal biomass (see figure 4.5). The concept of trees occupy-ing phytologically empty land may therefore be a reasonable approximation to reality in early Holocene Europe.

After two turbulent millennia of vegetation change, Euro-pean phytogeography looked fundamentally different. The boreal forest was pushed northwards to Scandinavia and north-ern Russia, tundra and steppe were all but removed from the scene, and the dominant vegetation type was now mixed de-ciduous forest. But if the distribution of plant formations had become essentially modern by 9000 Cal. yr BP, their composi-tion was not. This is immediately obvious from individual pollen diagrams, in which the characteristic feature is the con-tinued arrival and rise to dominance of new woodland taxa. After the pioneer woods of birch and pine, the first deciduous

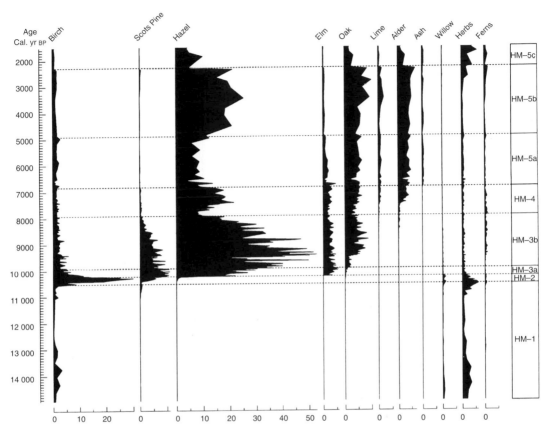

Figure 4.5 'Absolute' pollen diagram from Hockham Mere, Norfolk, England (modified after Bennett, 1983b)

trees to arrive in northwest Europe were hazel (*Corylus avellana*) and elm (*Ulmus*), both of which expanded without environmental constraints in less than 500 years (Bennett, 1983a). Later arrivals in the mesocratic forest were oak (*Quercus*), lime (*Tilia cordata*), alder (*Alnus glutinosa*) and ash (*Fraxinus excelsior*) (see table 3.1, p. 58). Other trees did not achieve their maximum extent until the declining or oligocratic stage of the Holocene, and some genera such as spruce (*Picea*) may still be expanding their range today (Huntley and Birks, 1983, p. 285).

Eastern North America

The return of temperate forest ecosystems began earlier in eastern North America, but in some areas finished later than it did in Europe. The principal reason for the early start was the terminal Pleistocene temperature rise which, in contrast to northwest Europe, was not broken by a major thermal oscillation in most areas (see chapter 3). Two dominant palaeoclimatic

Figure 4.6 Glacial to interglacial climatic changes in north Atlantic sector (based on COHMAP members, 1988); **A** environmental conditions as reconstructed from geological, pollen and ocean-core records; **B** numerical simulation models of past atmospheric circulation patterns

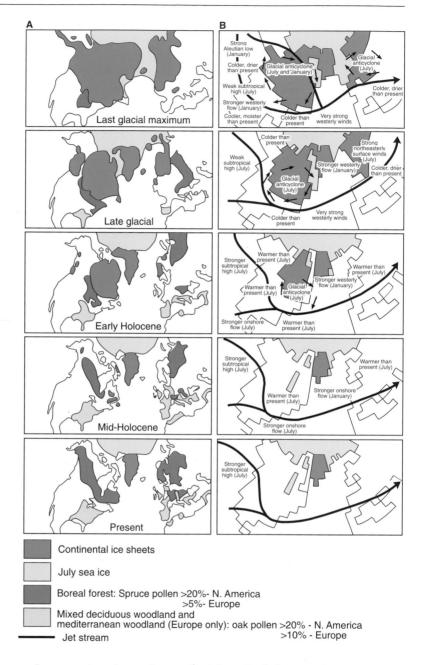

Continental ice sheets

July sea ice

Boreal forest: Spruce pollen >20%- N. America
                                    >5%- Europe

Mixed deciduous woodland and mediterranean woodland (Europe only): oak pollen >20% - N. America
                                                                                           >10% - Europe

——— Jet stream

and vegetational trends can be identified during the glacial-to-interglacial transition. The first was warming, which allowed temperate woodland species to spread northwards towards the retreating Laurentide ice sheet, pushing the boreal forest and tundra ahead of them. The second — desiccation — caused an eastward expansion of prairie grasses and forbs (leafy herbs). By the start of the Holocene, when Europe's vegetation still had a glacial character, the vegetation of the eastern United

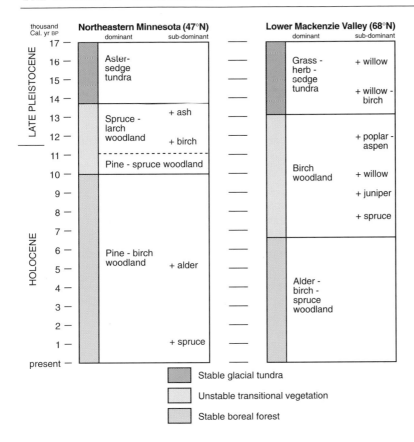

Figure 4.7 Pollen records of vegetation change at the southern and northern edges of North America's boreal forest zone (data from Ritchie, 1985; Cushing, 1967)

States already had a recognizably modern appearance (see figure 4.6A). To the north, however, the wasting mass of the Laurentide ice sheet stubbornly refused to move out of the way. At 11 500 Cal. yr BP ice, although much thinned, still covered most of what is now eastern Canada and it did not finally melt away until after 9000 Cal. yr BP. As a result, northern vegetation formations were compressed against the ice-sheet margins (Cushing, 1967; Webb et al., 1993). Trees grew right up to the edge of the ice, and led to the creation of buried forests, as at Two Creeks, Wisconsin, where the ice sheet temporarily readvanced over woodland around 13 800 Cal. yr BP.

The creation of the boreal forest that now covers eastern Canada consequently only took place during early Holocene times. The time-transgressive nature of the transition from glacial to interglacial ecologies is reflected in the timing of vegetation changes from site to site (Ritchie, 1987). At the southern margin of the boreal forest zone the unstable transitional period between tundra and forest lasted between c.14 000 and 10 000 Cal. yr BP, while at its northern limit it started c.13 200 and was not finished until after 7000 Cal. yr BP (see figure 4.7).

Dry mediterranean woodland

As with many arboreal vegetation formations, mediterranean ecosystems experienced a contraction in their range during glacial periods of the Pleistocene. This was most acute in the Mediterranean basin, where palynological research has yet to discover any clearly defined glacial refuge areas. Prior to *c*.14 000 Cal. yr BP, most of the northern littoral of the Mediterranean Sea appears to have been occupied by herb-dominated steppe in which *Artemisia* and chenopods were pre-eminent (Huntley and Birks, 1983; Rossignol-Strick, 1995). Evergreen oak (*Quercus ilex*-type) was continuously present in small amounts in the Sierra Nevada area of southern Spain between 30 000 and 14 500 Cal. yr BP (Pons and Reille, 1986). The Atlantic cedar and pistachio must also have had western Mediterranean refugia. The only area where evergreen oak, pistachio and olive are known to have survived together was the southern Levant (Palestine, Jordan), although even here they formed a relatively minor part of the overall plant cover during the late Pleistocene (van Zeist and Bottema, 1991; Horowitz and Gat, 1984).

The suggestion that the main glacial refuge for mediterranean flora lay in the eastern rather than in the western Mediterranean basin is given support by the earlier appearance of drought-adapted taxa pattern in the Holocene (see figure 6.10, p. 190). Where there was a westward expansion in tree species, it was by no means rapid, involving mean migration rates as low as 25 m per year in the case of manna ash (Huntley and Birks, 1983, p. 629). Furthermore, even after their appearance, typically mediterranean taxa usually represented a minor component of the flora (Reille and Pons, 1992). The herb-steppe which surrounded most of the Mediterranean Sea during the late Pleistocene was replaced during the early and mid-Holocene by sub-humid forest, sometimes dominated by conifers, more usually by broad-leaved deciduous trees (van Zeist and Bottema, 1991; Lamb and van der Kaars, 1995; Reille et al., 1996). The typically mediterranean formations of xeric evergreen forests, shrub and heathland are only rarely represented in early–mid-Holocene pollen diagrams, even at low altitudes such as the littoral zone of southern Crete (Bottema, 1980). When they are present they often indicate ephemeral plant associations, for instance at Fos-sur-Mer in southern France, where *Phillyrea*, pistachio and juniper occupied a geologically short-lived sand dune complex around 7000 Cal. yr BP (Pons and Quezel, 1985).

The summer-dry woodlands of California and Australia appear to have been much less severely restricted by glacial climatic conditions than those of the Mediterranean basin itself.

Fossil forests of calcified tree stumps and other ^{14}C-dated plant macrofossils indicate the existence of an ice-age refugium in coastal California (Johnson, 1977), while the coastal zone of southern Australia has supported a diversity of mediterranean-type habitats throughout the past 17 000 years (Chappell and Grindrod, 1983).

Tropical forests

Vegetation in the tropics was as strongly affected by glacial climatic change as vegetation in the temperate zone. In montane environments such as the Andes, reduced temperatures led to the altitudinally zoned vegetation belts being depressed by over 1500 m (see figure 3.3, p. 65). At the start of the Holocene as temperatures quickly rose, montane forest moved back up the mountains (Flenley, 1979b; Taylor, 1990). In lowland tropical environments it was moisture availability as much as temperature which controlled late Pleistocene and Holocene vegetation history (Bonnefille et al., 1990; Markgraf, 1993). During the period 20 000–15 000 Cal. yr BP, when active sand dunes extended into the present-day savanna zone, the low-land rainforests of Africa and South America were reduced to isolated pockets by the arid climate.

At present there are only a few lowland pollen sites with which to trace in broad outline the location of these glacial forest refuges or the pattern of post-glacial recolonization (e.g. Bush and Colinvaux, 1990). At the lowland site of Lake Valencia in Venezuela, the major transition recorded during the last 15 000 years occurred between 12 500 and 9700 Cal. yr BP (Bradbury et al., 1981). Before this, saline marshes occupied the lake basin and regional vegetation was dominated by open-habitat (i.e. non-tree) species. After it, Valencia was a deep freshwater lake, and the Holocene vegetation has remained one of savanna adjacent to the lake with evergreen and cloud forest on the surrounding mountains above 1000 m.

Lowland forest is unlikely to have responded instantaneously to climatic change. Moist forest elements returned early in the Holocene at some tropical sites, often abruptly as at Bosumtwi in Ghana, where the transition occurred around 10 000–9500 Cal. yr BP (Talbot et al., 1984), but later at others, such as Lake Euramoo in the Queensland rainforest of Australia (c.7400 Cal. yr BP). It is uncertain whether the African rainforest was more extensive during the early Holocene than in recent millennia. However, Alan Hamilton (1982) has suggested that the rise in pollen of *Podocarpus* (southern yew), which is widely recorded in East Africa around 4000 Cal. yr BP, marks a change from moister early–mid-Holocene woodland to drier late Holocene conditions.

Factors affecting forest re-advance

During the glacial-to-interglacial transition both individual taxa and whole plant formations shifted and usually expanded their ranges. But what were the factors that controlled the pattern of forest re-occupation of Europe, eastern North America and other regions at the onset of the Holocene?

CLIMATE was obviously the ultimate determinant of ecological change during the Quaternary ice ages (Delcourt and Delcourt, 1987; Huntley and Webb, 1989), and less severe glacial climatic conditions are probably the main reason for the greater diversity of tree taxa found in temperate North America than in Europe (Huntley, 1993a). However, it is by no means obvious that climate always directly controlled the rates of plant migration. In northern Europe at least, Coleoptera show an almost instantaneous rise in summer temperatures of around 7°C at the start of the Holocene (see figure 3.5, p. 75), a change confirmed by the early presence of thermophilous aquatic plants such as the reed-mace (*Typha latifolia*). It seems unlikely that tree distributions were therefore in equilibrium with climate during the first 1000–2000 years of the European Holocene (Birks, 1986a). Climate may have been a more important limiting factor in other regions, such as the eastern United States, where the temperature rise was probably less abrupt. Even here, however, climate cannot confidently be proposed as the dominant control in all instances, when one considers the manifestly unstable state of vegetation systems at that time.

RATES OF DISPERSAL AND MIGRATION of individual tree taxa are obviously likely to have played a paramount role once the brake of climatic control was removed. In this regard, species whose seeds are wind dispersed, such as birch and elm, have an obvious advantage over those like oak, which are reliant on other agencies, notably birds and streams, for carrying their acorns away from the parent tree. The growth rate and age of first seed setting for individual trees are other important determinants of the speed of migration (see table 3.1, p. 58). Mighty oaks from tiny acorns may grow, but they do so only slowly. The rates of migration of individual taxa at the start of the Holocene can be calculated from ^{14}C-dated pollen diagrams across Europe and eastern North America. These reveal that in Europe several opportunistic genera, notably birch, pine, alder and hazel, expanded at rates of 1–2 km per year over periods of 500–2000 years. That is, their seeds would have required dispersal over 10–20 km even if the trees were able to mature and fruit in only ten years (Huntley and Birks, 1983, p. 632).

In North America trees have maximum migration rates of around 0.5 km per year (Davis, 1976a), suggesting that other factors such as climate and competition with the existing spruce forest operated to slow down the return of deciduous trees. The American Jack pine (*Pinus banksiana*) provides an example. Today it is found in the northern Great Lakes region and in central Canada, but pollen maps show a glacial refuge in the southeastern United States, from which it spread northwestwards at an average rate of 350–500 m per year between 15 000 and 9000 Cal. yr BP (Davis, 1976a). Outlying populations of Jack pine still exist in New England as relics of its migration through the area 11 500 years ago.

Obviously the time of arrival of individual taxa depended not only on the rate of migration but also on the position from which they started their re-advance. In Europe and the Mediterranean basin, principal GLACIAL REFUGE AREAS have been identified in Iberia, Italy and southeast Europe (Huntley and Birks, 1983, p. 627; Bennett et al., 1991), to which can be added North Africa and the eastern Mediterranean. It was, for example, easier for pine to re-occupy northern Europe than it was for beech (*Fagus*), not only because of its faster rate of migration, but also because it started spreading from areas such as the formerly exposed continental shelf off western Ireland rather than from Italy and the Balkans. However, not all glacial vegetation refuges were as far removed from their modern distribution as they were in Europe or North America. In New Zealand, for example, southern beech (*Nothofagus*) was restricted to locally favourable sites in an otherwise open glacial landscape. With the rapid terminal Pleistocene warming that characterized most of the southern hemisphere, the forest re-established itself and within 300 years had replaced scrub and grassland (McGlone et al., 1993). This illustrates how rapidly plants can respond to climatic changes in the absence of soil or distance constraints.

Did lack of SOIL DEVELOPMENT restrict the re-advance of arboreal elements, especially of deep-rooting, nutrient-demanding species? In those areas formerly under ice, deglaciation left raw, undeveloped soils that favoured open vegetation (Pennington, 1986). In New England, for instance, sedge-dominated tundra is recorded at Rogers Lake between the retreat of the ice and *c.*13 500 Cal. yr BP, while at the same time spruce forest lay close by in the zone just outside the former glacial limits (Davis, 1976a; Watts, 1983). But although soil factors determined the speed and character of the early stages of plant succession around some ice-sheet margins, in other cases the arrival of new vegetation affected soil type more than the other way round (Willis et al., 1997). Soils may, in any case, have been less important in this protocratic phase than they were to be in the succeeding mesocratic and oligocratic periods. Regional soil differences, for example, were to influence strongly the polyclimax nature of Europe's mixed deciduous forest zone during the mid-Holocene.

COMPETITION between species certainly influenced early Holocene ecological succession. In northern Europe the rapidly dispersed silver birch (*Betula pendula*) prepared the ground for later arrivals with which it was then unable to compete. In particular it was shaded out by the dense forest canopy created by trees such as elm. The slower expansion of taxa such as oak and the failure of beech to establish itself as a major forest tree until the later Holocene may also have been due to competitive exclusion.

PHYSIOGRAPHIC BARRIERS such as the Alps and the North Sea hindered the free movement of forest taxa. On the other hand, the generally south to north drainage in Europe led to long-distance water dispersal of hazelnuts and acorns that would otherwise have spread more slowly. In eastern North America there are no major mountain ranges or epicontinental seas, although a major obstacle, no longer extant, did exist in the form of the Laurentide ice sheet. The effect of this on forest history can be seen clearly in the case of the spruce during the terminal Pleistocene. Rising temperatures allowed spruce forest to occupy much of the former periglacial zone of northeastern and midwestern United States. By the start of the Holocene, however, this boreal forest tree was being overtaken from behind by later invaders such as pine and oak, but was itself unable to move northwards on account of the residual ice mass occupying eastern Canada (see figure 4.6A). Squashed up against this physiographic barrier, spruce experienced a catastrophic population decline which is dated in pollen diagrams to around 11 500 Cal. yr BP and which, in some cases, occurred over as short a time as 50 years (Watts, 1983, p. 308). The ice that prevented forest invasion in North America ironically had the opposite effect in other parts of the world. The continued existence of the Laurentide ice sheet kept eustatic sea levels low and meant that, for example, the British Isles were not separated from the European mainland until after 9000 Cal. yr BP; that is, after the main wave of tree migrations was complete. In consequence, the native British flora lacks only oligocratic trees such as spruce (*Picea abies*). On the other hand, once isolated, island tree populations are vulnerable to local extinction, such as occurred in Ireland with pine during the late Holocene (Bradshaw and Browne, 1987).

Other factors, too, played their part. They include patterns of animal browsing, woodland management by hunter-fisher-gatherer groups, disease and fire. However, these were less important during the ecologically unstable transitional period than they were once the mesocratic forests had become established. It is also important to note that different factors operated over different spatial scales. Alder (*Alnus glutinosa*), for example, became established as a major European forest tree because of the existence of waterlogged soils at the base of local slope catenas, where it was able to compete most successfully. Finally, an almost universal characteristic of this period of dynamic vegetation change is that species behaviour was strongly individualistic (Delcourt and Delcourt, 1987). Particular genera or species may be associated with one another as part of a vegetation formation at the present day, but at times of climatic transition each one moved on its own, and some faster than others. Forest communities did not move *en bloc* at the start of the Holocene but were created and destroyed as individual taxa arrived and departed. This applies as much to tropical vegetation formations such as the montane rainforest of Papua New Guinea (Flenley, 1979a) as it does to the mixed deciduous forest of the temperate zone (Davis, 1976a). The variable migration rates and directions for different taxa created ephemeral plant communities which often lack a modern analogue – such as the combination of spruce, oak and ash which existed in the midwestern United States at the end of the Pleistocene and which do not now occur sympatrically (Watts, 1983).

The Ecology of Mesolithic Europe

During the late Pleistocene, Europe's mid-latitude tundra had supported large herds of reindeer (*Rangifer*), horse (*Equus*) and other herbivores (see chapter 3). This fauna was displaced northwards at the start of the Holocene as temperate forests returned, and in its place came new woodland animals such as red deer (*Cervus elaphus*), aurochs (*Bos primigenius*) and wild boar (*Sus scrofa*). As resources for human exploitation, these animals were more dispersed and less visible in the forests than had been the concentrated and easily culled fauna of the Late-glacial tundra. But if animal resources became less accessible after *c.*11 500 Cal. yr BP, plant resources changed in exactly the opposite way. Mixed deciduous woodland is a biome of high, if strongly seasonal, primary productivity and contains several hundred potentially edible plant species. These range from tree products such as hazel nuts through berries and fruit to fungi and bracken rhizomes. And, as David Clarke (1976) reminded us, these could be collected for much less expenditure of effort than their hunted meat equivalent. Any child over four could gather sufficient to feed itself, and in habitats like the stone pinewoods of southern Europe it would collect enough for the rest of the family too (see plate 4.3). The

Plate 4.3 The Mediterranean stone pine (*Pinus pinea*); its kernels have a protein value two-thirds that of lean steak

consumption of plant foods generally leaves fewer traces than the bones left after animal hunting, and it is consequently difficult to make quantitative estimates of the relative dietary importance of each at individual archaeological sites. None the less, the fact that bone remains are well represented at Mesolithic sites indicates that Europe's early–mid-Holocene population was omnivorous rather than vegetarian.

The seasonal rhythm of plant growth and animal movement in temperate woodland ecosystems was strongly to influence the food schedules and lifestyles of Mesolithic h-f-g groups. In northern England, for example, Mesolithic sites are found in two clear elevational bands: one within 100 m of modern sea level, the other on high ground between 250 and 500 m (Jacobi et al., 1976). Among the lowland sites is Star Carr, the scene of famous archaeological investigations by Grahame Clark between 1949 and 1953 (Clark, 1972). His were among the first excavations to be as concerned with the site economy and environment, as revealed by bones, seeds and pollen grains, as with the typology of stone tools. Study of elk (*Alces*) and red deer antlers found at Star Carr initially suggested that the site was most intensively occupied in winter and early spring. Clark proposed that in summer most of the inhabitants moved away with the red deer to the higher ground of the North Yorkshire Moors, which were then covered by open birchwood and

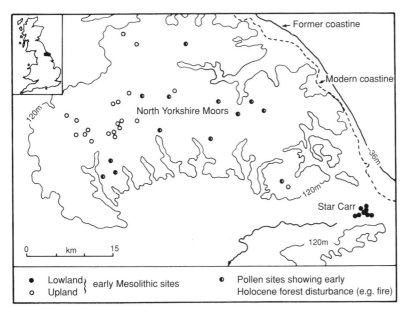

Figure 4.8 Pattern of Mesolithic settlement and early Holocene vegetation disturbance in North Yorkshire, England (based on Jacobi, 1978; Simmons and Innes, 1985)

hazel scrub. This seasonal transhumant cycle would, he suggested, have allowed deer and their hunters to avoid the winter snows on the high ground and the insect-ridden lowlands in summer. Subsequent researchers (e.g. Caulfield, 1978; Andresen et al., 1981) have cast doubt on Clark's original model of Star Carr as a winter base camp; the antlers, it is thought, were brought onto site by hunters rather than being shed by the deer locally. Re-analysis of the animal bones by Tony Legge and Peter Rowley-Conwy (1988) has instead led to the interpretation of Star Carr itself as a late spring–summer hunting camp. None the less, it is likely Mesolithic bands followed a seasonal cycle involving 'fission and fusion' similar to that observed in modern hunters and gatherers (Mellars, 1976). Bands would probably have come together at lowland base camps to pass the winter – a season of climatic stress and food shortage – but split up into smaller family-sized units across their summer territory. This model is given archaeological support from site types, lowland Mesolithic sites being generally larger and more substantial (*c.*20 people) than their upland equivalents (*c.*5 people) (see figure 4.8).

As the relatively open birch, hazel and pine woods of the protocratic phase gave way to the closed-canopy mesocratic forest, so open habitats became increasingly scarce. In lowland environments openings came to be found mainly near to water, either on the coast or next to lakes, marshes and rivers. Star Carr itself comprised a birchwood platform at the swampy edge of a lake, now infilled (Cloutman, 1988); the waterlogged conditions being responsible for the excellent preservation of

organic remains at the site. Wetlands are resource-rich environments, whether for edible water-plants such as cress and
water-lily (Clarke, 1976), or for fowling and fishing. The latter
was especially important in Ireland, which had been cut off by
rising sea levels before many woodland animals had time to
arrive (see table 4.1). Compensation for Ireland's depauperate
mammal fauna came from that island's many lakes and rivers.
For example, along the Lower Bann river between Lough Neagh
and the sea, the bones of large numbers of migratory fish,
notably salmon and eel, have been recovered from Mesolithic
excavations (Woodman, 1978, 1985).

In upland environments, openings in the forest lay mainly
towards the upper altitudinal limit of tree growth. Here fire
provided an alternative method of deforestation to ring-
barking or felling with Mesolithic tranchet stone axes. Selective
burning is a traditional technique of environmental management practised by hunters and pastoralists throughout the
world (Mellars, 1975). The new vegetation growth after a fire
increases grazing and browsing potential, and the number of
deer or cattle that can be supported responds accordingly.
Charcoal provides one of the best palaeoecological indications
of past fire frequencies (Patterson et al., 1987). Charcoal in soil
and peat profiles, and palynological evidence for a reduced
forest cover in areas such as the southern Pennines and Dartmoor, suggest that recurrent burning of upland vegetation took
place during the later Mesolithic (Simmons and Innes, 1985;
Tallis and Switsur, 1990; Caseldine and Hatton, 1993). It is
probable that much of this resulted from deliberate firing,
as this ecotone experienced substantial and long-continued
human occupation during the Mesolithic period.

The Early Holocene in the Tropics

For many years the tropics were a backwater of Quaternary
research. Wetter and drier climatic phases were recognized,
it is true, but these were fitted around the glacial–interglacial
frameworks developed in higher latitudes, notably in the Alps.
Then, around 1960, matters began to change. One reason for
that change was the discovery at Olduvai Gorge in Tanzania of
hominid bones dated radiometrically to over 1.5 million years
old, indicating that the human race had its origins on the
African continent. A second change, more directly relevant to
this chapter, came with the application of ^{14}C dating to fossil
lake beds in the Sahel and the East African rift (see plate 4.4).
These former lakes, interpreted as the product of a wetter
'pluvial' climate, had been correlated with the glacial phases
of Europe and North America and might have been expected

Species	Late-glacial	Early Holocene		Present day	
		Britain	Ireland	Britain	Ireland
bison	x				
cave bear	x				
giant deer[a]	x				
mammoth	x				
woolly rhinoceros	x				
lion	x				
lemming	x				
arctic fox	x				
tundra vole	x				
elk	x	x			
horse	x	x			
reindeer	x	x			
wolf	x	x	x		
brown bear	x	x	x		
lynx	x	x	x		
otter	x	x	x	x	x
stoat	x	x	x	x	x
blue hare	x	x	x	x	x
pygmy shrew	x	x	x	x	x
fox	x	x	x	x	x
badger		x	x	x	x
pine marten		x	x	x	x
red deer		x	x	x	x
wood mouse		x	x	x	x
wild cat		x	x	x	
polecat		x	x	x	
wild boar		x	x		
common shrew		x		x	x
mole		x		x	x
hedgehog		x		x	x
red squirrel		x		x	x
brown hare		x		x	x
roe deer		x		x	
weasel		x		x	
field and water vole		x		x	
yellow-necked mouse		x		x	
water shrew		x		x	
aurochs		x			
beaver		x			
rabbit				x	x

[a] Megaloceros

Source: Grigson, in Simmons and Tooley (1981) and other sources

Table 4.1 Terrestrial mammals of the British Isles during the terminal Pleistocene and Holocene

Plate 4.4 Early Holocene lake sediments (10 m thick) from Taoudenni in the central Sahara, subsequently eroded by the wind to form pillars

to produce calibrated ^{14}C ages between 22 000 and 12 000 years ago. However, when the French geologist Hugue Faure sent off samples of freshwater carbonate and diatomite from lake basins in the Sahel, he received back much younger ^{14}C ages, between 10 300 and 7700 years old after calibration (Faure et al., 1963). Not surprisingly this encouraged further research work, which similarly yielded ^{14}C dates in the first half of the Holocene. By 1972 Karl Butzer and others were able to put together the results from a number of East African lakes and show that their water levels had fluctuated in concert over the past 15 000 years. Non-outlet lakes were clearly an important source of palaeoclimatic information in low-latitude regions. By the 1970s, so many lake basins were under study and so many ^{14}C dates produced that it was necessary to centralize them in a global databank, compiled by Alayne Street-Perrott at Oxford, and now housed at Lund University in Sweden. In this database lakes were classified as at high, intermediate or low levels at 1000-year intervals within the time period for which reliable ^{14}C chronologies exist (Street and Grove, 1979; Street-Perrott et al., 1989). From before 10 000 to c.5500 Cal. yr BP there was a truly massive extension in the zone of high-level lakes and water surplus, not only into the semi-arid Sahel but into the Sahara desert itself (Fontes and Gasse, 1991).

Saharan palaeoecology

Until almost the time of Egypt's pyramid builders (4550 Cal. yr BP) the ecology of the Sahara was not the present hyper-arid wasteland but a veritable land of lakes. The presence of permanent freshwater is indicated not only from mollusc shells, diatoms and laminated lake sediments but also – and more spectacularly – from bones of large aquatic fauna. From northern Mali, where rainfall now averages only 5 mm per year, have come finds of crocodile (*Crocodylus niloticus*) up to 3 m long and hippo (*Hippopotamus* sp.) (Petit-Maire and Riser, 1983). These and other fauna typical of savanna ecosystems required not only water but also plant life in order to exist.

Pollen grains are generally poorly preserved in arid environments but several sites covering the full width of the Sahara have now been successfully investigated. In the west (Senegal, Mauritania), relict forests of the humid Guinean type confirm pollen evidence of a southerly vegetation shift in vegetation zones since mid-Holocene times (Lézine, 1989). A strong mid-Holocene augmentation in pollen of sub-humid Sudanian vegetation is also recorded in the Lake Chad basin, now in the semi-arid Sahelian climatic and vegetation zone (Maley, 1981). But perhaps the most startling evidence has come from a site named Oyo in the heart of the hyper-arid eastern Sahara (Ritchie

Plate 4.5 Mid-Holocene rock painting of a giraffe head from Tassili in the arid heart of the Sahara desert. Under the former wetter climate, savanna wildlife was common and so, too, were people hunting them

and Haynes, 1987). In spectra dating from before 9500 until
*c.*7000 Cal. yr BP, most non-local pollen was of Sudanian type,
including grains of tropical taxa such as *Hibiscus*, which are
produced by plants in only small quantities and are poorly
dispersed (see figure 4.9). During this period Oyo – then a
deep, stratified lake – must have been surrounded by savanna
vegetation similar to that now found 500 km to the south.
Between 7000 and 5700 Cal. yr BP the lake became shallower
and acacia-thorn and then scrub grassland replaced the sub-
humid savanna. Finally, around 5100 Cal. yr BP the lake dried
out, aeolian activity recommenced and vegetation disappeared
except in wadis and oases.

The Saharan desert, which had been considerably more
extensive than at present during the late Pleistocene (20 000–
15 000 Cal. yr BP), effectively did not exist during most of the
early Holocene (10 000–5500 Cal. yr BP). The savanna that replaced most of the present arid zone was relatively hospitable
to human life and there is abundant archaeological evidence,
including some superb rock paintings, that the human ecology
was then radically different from that which could exist in the
Sahara today (see plate 4.5). Of course, the environmental resources varied considerably across this huge area. In the Egyptian Sahara, for example, early Holocene game was dominated
by gazelle (*Gazella dorcus*) and hare (*Lepus capensis*), and
lacked the diversity of the then contemporary savanna fauna
of northern Mali (Gautier, 1980). For the most part human
subsistence patterns varied according to the resources available,
but the variety of regional specialities would have impressed
even the best French chef. Freshwater turtles, molluscs and
fish were an important component of diet in the southern
Saharan Neolithic, which has been appropriately labelled the
Aqualithic culture. On the northwestern edge of the Sahara,
where conditions were semi-arid rather than humid, the distinguishing feature of the cuisine was its fondness for escar-
gots – edible land snails. The so-called Escargotières of the
Capsian culture are small shell middens that were occupied
occasionally but repeatedly for up to two or three millennia
during the early Holocene (Lubell et al., 1976). Each of the
snail species exploited occupies a separate ecological niche
that has a somewhat different seasonal ethology so that foraging would have rotated from site to site, allowing the snail
population to recover in between times, preventing over-
predation. Although whole shells averaged 25 000 per cubic
metre at excavated sites, the image of a population feasted on
this culinary delicacy is probably a false one. In truth, the snails
were probably placed in the pot along with meat and wild
plants, and on their own accounted for only a limited proportion of the total dietary base.

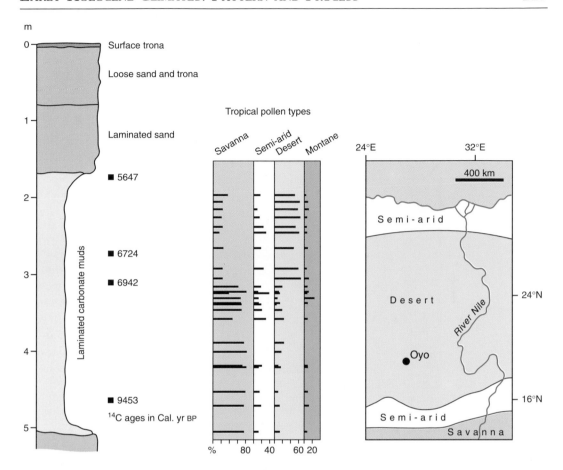

Figure 4.9 Pollen diagram from Oyo, eastern Sahara (after Ritchie et al., 1985)

Both of these economies fall within the province of hunting-fishing-gathering, but a third adaptation to Holocene Saharan environments involved a more substantial manipulation of food resources. In what is now the Egyptian desert, Fred Wendorf and his colleagues (1976) have discovered and excavated Neolithic sites which included some domestic species, notably cattle. The beef-eaters of this area lay adjacent to the major Near Eastern nuclear hearth of plant and animal domestication and they probably represent the origin of the cattle-based cultures which were to spread throughout Africa during the course of the Holocene.

Early Holocene Climates: Pattern and Process

It is fair to suggest that our perception of Holocene climatic changes has had a Eurocentric origin. In temperate Europe it has not been easy from pollen or other proxy data to identify unambiguously any major climatic shifts since the last glaciation ended. Certainly cooler/wetter or warmer/drier phases of

TECHNICAL BOX

Technical Box V: The Blytt–Sernander classification of the European Holocene

Vegetational and inferred climatic changes have provided the main basis subdividing the European Holocene. One especially influential framework has been the Blytt–Sernander model, originally based upon peat stratigraphy. Phases of faster and slower peat growth were used to divide the Holocene into phases of inferred wetter and drier climate (see table 4.2). With the subsequent development of palynology, pollen zones were fitted in with the same scheme.

Table 4.2   Blytt–Sernander stages for the European Holocene

Period	Peat type	Inferred climate	Pollen zone	Approximate age Cal. yr BP
Pre-Boreal	hydrosere peat	cool/dry	IV	11 500–10 500
Boreal	humified peat + pine stumps	warm/dry	V/VI	10 500–7800
Atlantic	unhumified peat	warm/wet	VIIa	7800–5700
Sub-Boreal	humified peat + pine stumps	warm/dry	VIIb	5700–2600
Sub-Atlantic	unhumified peat	cool/wet	VIII	2600–present

While recent work has supported the idea that phases of peat growth were caused by climatic changes, deficiencies have been found in the original scheme. Most ombrotrophic bogs in fact show a more complex sequence of recurrent surfaces from humified to unhumified peat than were recorded by Blytt and Sernander (see figure 6.2, p. 163). Furthermore, with the application of ^{14}C dating it has become clear that Holocene vegetational changes were time-transgressive (Mangerud et al., 1982). The Blytt–Sernander scheme is consequently falling into disuse, although in a colloquial sense terms such as the 'Atlantic oakwoods' are still used.

the Holocene were recognized by Blytt and Sernander (see Technical Box V). Similarly, a thermal optimum (9000–5500 Cal. yr BP) was identified by Iversen and others from the former extension of plants such as the water chestnut (*Trapa natans*) and animals such as the pond tortoise (*Emys orbicularis*) north of their present climatic limits. But these changes appeared minor in comparison with the fluctuations in climate that affected Europe during the millennia prior to 11 500 Cal. yr BP. It was therefore presumed that 'from at least 7000 BC, and possibly earlier, the worldwide climate had been essentially the same as that of today' (Raikes, 1967).

It is now known, however, that elsewhere in the world climatic stability did not exist during the Holocene. In the tropics and sub-tropics, in particular, the climate 9000 or 6000 years ago was substantially different from that of the present day (COHMAP members, 1988). These differences are clearly manifested in palaeoecological and geological records and can only realistically be explained by shifts in atmospheric circulation patterns. This should cause us to re-examine more closely the climatic record of the temperate zone. Was the climate of Holocene Europe really as stable as has been often envisaged? Or was the dominant vegetation formation of mixed deciduous forest simply ecologically insensitive to climate?

Of course, climatic interpretation of proxy data is not a straightforward process. Former high lake levels, for example, could be the product of reduced evaporation at times of low temperatures rather than any absolute increase in precipitation. Fortunately, there is no such problem of interpretation with the enlarged early Holocene lakes of the tropics. They formed when the climate may even have been slightly warmer than that of the present day, so high water levels strongly imply higher rainfall. Calculations using both water and energy balance approaches indicate that rainfall in East Africa and the Sahara increased by between 150 and 400 mm per annum (Street-Perrott et al., 1991).

What was the source of this enhanced moisture? Here, too, the palaeoenvironmental record is fortunately clear in indicating that the zone of tropical convectional rains moved northwards and broadened during the early Holocene. Mid-latitude depressions may have interacted with tropical air masses over some parts of the western Sahara and over high ground such as the Tibesti Plateau, but were of only secondary importance (Nicholson and Flohn, 1980). A northerly shift in the ITCZ (Inter Tropical Convergence Zone) and its associated rains is indicated by a poleward shift in the Sudanian-Sahelian vegetation belts, and by Holocene lakes that were larger, deeper and more permanent towards the southern side of the Sahara.

A still more important indicator comes from changes recorded outside Africa. If the ITCZ moved north over the African continent it would also have been expected to do so over the Indian Ocean sector, which today forms a linked part of the same monsoonal circulation system. And indeed, comparable evidence for enhanced summer rainfall occurs in the Arabian and Rajasthan (northwest Indian) deserts, in the form of fossil lake beds (McClure, 1976), palaeosols and pollen indicative of savanna vegetation (Singh et al., 1974). Indian Ocean sediment cores also record stronger upwelling of the Arabian coast and pollen of tropical origin (van Campo et al., 1982; Sirocko et al., 1991). There is even evidence of high lake levels and changed vegetation composition as far north as the plateau of Tibet (Gasse and van Campo, 1994). This suite of palaeoclimatic data leaves little doubt that the whole Asian monsoonal system was strengthened during early Holocene times. Similarly, pollen and lake-level sequences showing wetter early–mid-Holocene climates have been found in other parts of the tropics, such as the Caribbean and northern Australia (Leyden, 1987; Hodell et al., 1991; Schulmeister, 1992).

Increased summer rainfall may also have been responsible for the dominance of sub-humid forest over most of the Mediterranean basin during the first half of the Holocene. The elimination of a clear dry season would have disadvantaged drought-adapted forms such as the olive, and would also have greatly reduced the risk of fire, which has otherwise been an important agent in the maintenance of dry sclerophyll shrub and woodland. Enhanced early–mid-Holocene precipitation is recorded in other mediterranean-type regions, such as south Australia, in the form of higher-level freshwater lakes and – significantly – from grass pollen indicative of enhanced summer monsoonal rainfall (Bowler, Singh in Chappell and Grindrod, 1983; Luly, 1993). In fact, precipitation change in mediterranean latitudes becomes eminently logical when viewed in the context of the poleward movement of monsoon-type rain into sub-tropical deserts such as the Sahara. Increases in rainfall may have been less marked than in the tropics (+20 per cent has been suggested in south Australia), but on account of their seasonal distribution they had a critical effect on plant ecology.

However, shifts in atmospheric circulation during the early–mid-Holocene did not bring increased rainfall to all mid-latitude regions. Those areas out of reach of sub-tropical moisture sources instead became drier than they are at present. In eastern Turkey and western Iran, for example, the re-advance of woodland vegetation commenced around 12 500 Cal. yr BP, but was not complete until 6300 Cal. yr BP (see figure 5.7, p. 145), and this time lag may have been linked to early Holocene climatic aridity (Roberts and Wright, 1993). An even clearer pattern occurred on North American prairie between the Rocky Mountains and the Mississippi. The mid-western prairie–forest boundary moved eastwards until 8000 Cal. yr BP, but subsequently shifted back in the opposite direction (Webb et al., 1984). As this boundary is today controlled by moisture availability it is reasonable to use it as a proxy climatic indicator, recording climatic conditions between 8000 and 4500 Cal.

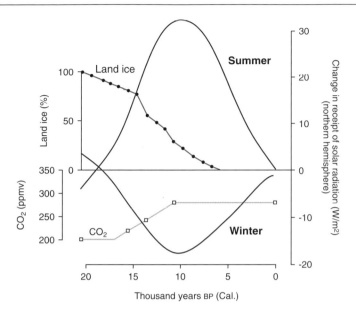

Figure 4.10 Changing receipt of solar radiation and other climatic controls in the northern hemisphere since the last glacial maximum (modified after Kutzbach and Street-Perrott, 1985)

yr BP which were drier than at present. Sediment cores from some of the many small lakes of the North American Great Plains also record this climatic phase. Diatom and geochemical analyses from Devil's Lake in North Dakota, for example, show high salinities during the first half of the Holocene, while before and after this phase the lake water was relatively fresh (Fritz et al., 1991; Haskell et al., 1996).

A sequence of vegetation revertence is also evident further north in America at the boundary of boreal forest and tundra (Ritchie, 1984). Between 13 000 and 8000 Cal. yr BP Canada's boreal spruce forest moved northwards, only to retreat southwards after 4500 Cal. yr BP (see figure 4.6). In the highlands of New Guinea and South America too, vegetation belts lay 100–200 m above modern elevations between c.10 000 Cal. yr BP and the mid–late Holocene (Flenley, 1979b). Climatically calibrated pollen data for Europe also show most of Europe having higher summer temperatures than at present (Huntley and Prentice, 1988). These and other lines of evidence suggest temperatures slightly higher than those of today – supporting Iversen's idea of a mid-Holocene 'thermal optimum'. However, some regions, like the Mediterranean basin and parts of the tropics, may have been cooler, possibly linked to increased cloud cover and soil moisture. Any warming may also have been restricted to the summer half of the year.

The cause of these climatic changes probably lies in astronomically controlled variations in the Earth's receipt of solar radiation (see figure 4.10). At the start of the Holocene one of the three Croll–Milankovitch cycles – the precession of the equinoxes – lay at the opposite point on the cycle from that of

Table 4.3 Mean annual hydrological changes between 9000 Cal. yr BP and the present day

	Global		Northern hemisphere only	
	Land (%)	Ocean (%)	Land (%)	Ocean (%)
Evaporation (E)	+1.3	−2.4	+2.1	−2.6
Precipitation (P)	+9.7	−6.0	+12.0	−4.0
P-E	+8.5	−3.6	+9.9	−1.4

Positive values record an increase at 9000 Cal. yr BP compared to the present

Source: GCM experiment of Kutzbach and Gallimore (1988)

the present day. This meant that the northern hemisphere then received nearly 8 per cent more solar radiation during summer months than it does at present. Northern summers were considerably warmer and heating over the land, especially over south Asia, caused low pressure which sucked in the humid, monsoonal air masses from the south. Numerical models of the Earth's atmospheric circulation simulate this strengthened monsoon and an increase in effective precipitation over northern hemisphere land areas (see table 4.3). These models also show good agreement with Holocene palaeoclimatic data for Africa and south Asia (Kutzbach and Street-Perrott, 1985; COHMAP, 1988; Wright et al., 1993). This orbital 'forcing' of Holocene climate appears to have been amplified by the biosphere itself. Modelling experiments have shown that vegetation had a significant effect on climatic conditions in key boundary areas, such as the edges of forest and desert biomes, by changing the surface albedo and moisture conditions. Along the boreal–tundra ecotone, for example, the expanded forest has been calculated to have itself caused an additional warming of about 4°C in spring temperatures 6000–7000 years ago, over and above that caused directly by solar heating (Foley et al., 1994). The Saharan and Arabian deserts, too, may have shrunk by more than expected in the first half of the Holocene because of feedback effects (Street-Perrott et al., 1991; Kutzbach et al., 1996); it seems that once the desert was transformed to savanna it had an inherent tendency to stay that way.

The climate during the early–mid-Holocene 'thermal optimum' has been used to test and calibrate numerical models of the atmosphere, which are also used to simulate what the Earth's climate may be like if the 'greenhouse' effect causes a general increase in temperatures during the twenty-first century (Wright et al., 1993). On the other hand, the Holocene does not seem likely to offer a close analogue for a future greenhouse-gas-warmed climate because it was dominated by seasonal, rather than mean annual, changes in the Earth's net receipt of solar radiation (Mitchell, 1990). Holocene climatic changes also offer a warning about the hazards of climate prediction. On at least three occasions the Holocene climate deteriorated abruptly, only to return to its previous state within a few centuries (see figure 4.11). Pierre Rognon (1983) termed these events 'climatic crises'. They are recorded in abrupt falls in the levels of many of Africa's lakes, dated to c.12 000 Cal. yr BP, 8200 Cal. yr BP, and 5200 Cal. yr BP (Street-Perrott and Roberts, 1982; Gasse and van Campo, 1994). As with the rapid rises in temperature at the Pleistocene–Holocene transition, these hydrological changes occurred so rapidly that their true rate cannot be determined by ^{14}C dating. The ecological consequences of these arid episodes were sometimes catastrophic; for instance at Lake Shala in Ethiopia, one lake recession resulted in the death of fish on a massive scale (Gillespie et al., 1983). On the other hand, pollen analysis indicates that trees – being long-lived – were, in some cases, able to survive through the extended droughts (Lamb et al., 1995). Importantly, the abrupt events are also recorded by the Greenland ice-core record as sharp troughs in atmospheric methane, implying a reduction in tropical wetlands which are a major source of 'marsh gas' (see figure 4.11). These droughts must have resulted from short-lived mechanisms such as the switching off of the Atlantic Ocean 'conveyor' (see chapter 3) rather than long-term changes in radiation input, and they not surprisingly create considerable uncertainty for predicting future climates.

Conclusion

The key theme of the early part of the Holocene is one of adjustment and adaptation to new environmental conditions. Eustatic sea level adjusted to the reduced global volume of ice, and in turn modified river base levels and coastal ecologies. Soil–plant associations responded to climatic warming by dramatically changing their extent and distribution. The world's deserts and tundras became much less extensive than they had been under a glacial regime, while temperate grasslands expanded, at least during the first half of the Holocene. Above all, forest ecosystems – both temperate and tropical – were greatly enlarged as higher temperatures and moisture encouraged tree growth, so increasing global NPP and biomass (see table 3.3, p. 63). All these ecosystems brought with them a host of new insects, birds and mammals. On the expanding North American prairie, for example, bison were presented with a new and largely vacant ecological niche in which they were able to increase their populations enormously. But biomes

Figure 4.11 Two tropical lake-level records compared to atmospheric methane concentration recorded in a Greenland ice core (data from Gillespie et al., 1981; Talbot and Delibrias, 1977; Blunier et al., 1995). Arrows show abrupt arid events interrupting the longer-term climatic trends. The map shows the extent of high lake levels in Africa during the early Holocene

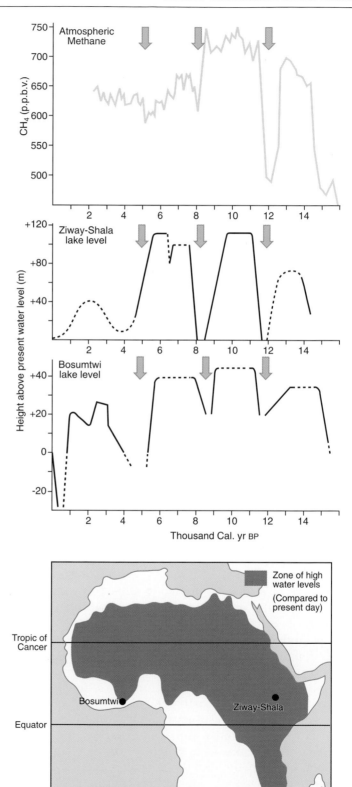

did not move from their glacial to their interglacial positions *en masse* like a well-disciplined army. Rather, different species set off from different places and arrived at different times, like the anarchic race at the Mad Hatter's tea party in *Alice in Wonderland*. In other words, migrations and adaptations were achieved by individual taxa rather than by whole plant formations.

It was not only physical and biological regimes that had to adapt; so too did cultural ones. The advent of the Holocene brought radically altered environmental resources to the world's human population. Mammalian extinctions and the reduction in open habitats such as tundra meant a reduced dependence on meat and other hunted animal products. There was also habitat loss in the coastal zone as sea levels rose. By contrast, forests and wetlands increased, and it is no surprise that from the European Mesolithic to the Saharan Aqualithic, early Holocene habitations are found on what were then lake shores (Sauer, 1948).

It was believed by classical authors such as Varro and Seneca that there had once been a 'Golden Age', 'when man lived on those things which the virgin earth produced spontaneously' and when 'the very soil was more fertile and productive' (cited in Glacken, 1967, p. 133). If ever there was such a 'Golden Age' then surely it was in the early Holocene, when soils were still unweathered and uneroded, and when Mesolithic peoples lived off the fruits of the land without the physical toil of grinding labour (Sahlins, 1974). Human life in the Mesolithic was adjusted to the movement of deer or salmon, the autumn harvest of fruits, nuts and berries, and other rhythms of nature in the same way it was in h-f-g groups like the American Kwakiutl Indians who were encountered by Europeans in the eighteenth and nineteenth centuries AD.

But already the human role in nature was far from passive. Although there was little benefit to be gained from cutting down trees – as it was the forest that provided the resources necessary for human life – these groups were well capable of manipulating those plant and animal resources of use to them. There is an unusually high concentration of ivy (*Hedera helix*) pollen at many European Mesolithic sites, for example; the most likely explanation is that ivy was used to attract red deer (Simmons and Dimbleby, 1974). Similarly hazel, which coppices freely in response to burning, is much more abundant in the protocratic phase of the Holocene than in any previous Pleistocene interglacial – possibly an indirect result of Mesolithic use of fire (Smith, 1970; Huntley, 1993b). The use of fire and other cultural impacts meant that the 'wildwood' of the early–mid-Holocene may not have been in quite the 'virginal' state Varro imagined.

Even more portentous were changes that were taking place in the subsistence base. Although the dominant mode of production remained founded in seasonally mobile h-f-g economies, the food base became broader and more diversified. In a few environments, such as the basin of Mexico, it proved possible to schedule hunting and collecting of seasonal resources in such a way that foodstuffs were available in one place all year round (Niederberger, 1979). A related form of early Holocene adaptation involved intensive use of certain non-tree food plants, particularly grasses. The combination of harvesting wild grass seeds and a sedentary existence was to lead to an economy fundamentally different from all that had gone before; one that was irreversibly to change human relations with nature. That economy was farming.

CHAPTER FIVE

The First Farmers

Agricultural Origins

In the previous two chapters attention was focused on the transition from a glacial to an interglacial world, and the consequences of climatic change for natural and cultural ecologies. In this chapter we turn to humans as agents of environmental modification; more specifically, to the advent of agricultural modes of production. These in turn were made possible by the domestication of plants and animals.

The question of where and when our domestic crops and animals originated is a truly interdisciplinary problem which has received contributions from – amongst others – geographers such as Carl Sauer (1952), archaeologists such as Eric Higgs (1972, 1975) and David Harris (Harris and Hillman, 1989; Harris, 1996a), and crop scientists such as Jack Harlan (1975, 1995). However, our story starts with the Russian plant geneticist Nicolai Vavilov, who in 1926 proposed eight hearth areas as the most likely centres for the origin of individual crop complexes (Harris, 1990). Most of these lay in a belt across the southern, sub-tropical margins of Asia and Europe (see figure 5.1). Vavilov based his hearths on areas of maximum genetic diversity, which he thought would include the nearest wild relatives to plant domesticates. This, of course, assumed that the plant distributions had not changed significantly since the time of their domestication, for instance as a result of climatic change. Also underlying Vavilov's work was the belief that domestication of particular forms occurred only at specific

Figure 5.1 Vavilov's proposed hearth areas of plant domestication (based on Harlan, 1971)

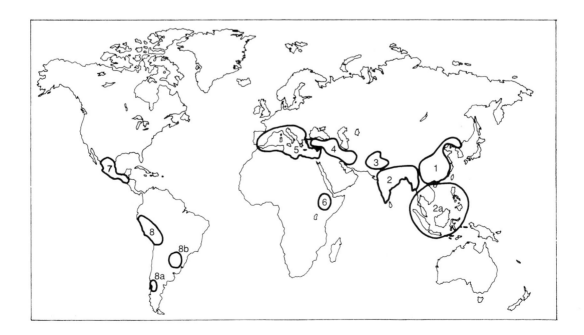

times and places, rather than being a continuing process that was diffuse in time and space. The present, much wider distribution of domestic crops was thought to be a result of subsequent diffusion from the original hearths into adjacent cultures.

Vavilov narrowed the field of possible locations for plant domestication, but he was unable to show that it had actually occurred where he suggested, nor did he have any firm ideas about when. This dearth of historical information has only been countered since 1950 as a result of collaboration between archaeologists on the one hand, and botanists and zoologists on the other. The latter have studied the animal bones and charred or waterlogged seed remains from excavated archaeological sites. Integrated bio-archaeological research has produced some of its most successful results in the Near East (Vavilov hearth 4) and Mesoamerica (hearth 7).

Near East

The Near East is probably the earliest and arguably the most important of all the global centres of domestication. From the Neolithic farmers of this region we have inherited cereal crops such as wheat (*Triticum*), barley (*Hordeum*), rye (*Secale*), pulses including pea (*Pisum*) and lentil (*Lens*), and a range of domestic animals including cattle (*Bos*), sheep (*Ovis*), goat (*Capra*), pig (*Sus*) and dog (*Canis*). The increased economic productivity and new social organization produced by this 'Neolithic revolution' at the beginning of the Holocene was associated with population increase and great developments in material culture.

Integrated bio-archaeological field research into the origins of agriculture was initiated in the 1950s by American Robert Braidwood. Excavations of mounds – or 'tells' – such as Jarmo and Jericho revealed Neolithic villages ^{14}C dated back to the very start of the Holocene (figure 5.2). Their mudbrick dwellings contained charred seeds, whose subsequent analysis, notably by Danish archaeobotanist Hans Helbaek, showed many of them to be early domestic cereals significantly different from naturally occurring wild varieties (see Technical Box VI). A big advantage of this work is that it could be compared against a rich modern ethnobotanical record, for traditional agriculture remains little changed in some parts of the Near East from that which existed during Neolithic times (Hillman, 1984) (see plate 5.1). The identification of domestic varieties was less easy with animal bones, but criteria such as size and age–sex structure of the animal population indicated that these were herded rather than hunted (Clutton-Brock, 1987; Meadow, 1989). The success of this work in turn encouraged further research and in consequence there is now a relatively detailed record of the cultural and economic prehistory of the Near East between about 15 000 and 8000 Cal. yr BP (i.e. terminal Pleistocene and earliest Holocene) based on bio-archaeological data. Although some areas such as Afghanistan are still poorly known, enough information now exists to suggest with some confidence when and where cereal-based farming began (Moore, 1985; Zohary and Hopf, 1993).

Archaeological sites older than *c.*12 000 Cal. yr BP belong to Mesolithic-type cultures, such as the Natufian culture of Palestine (Bar-Yosef and Belfer-Cohen, 1992). Sites were relatively small and most were only occupied seasonally. Many site economies included harvesting of wild cereal stands, while others involved husbandry of essentially wild herds of sheep and goat – but in both cases these formed part of broadly based economies utilizing a wide range of resources (Hillman, 1996). During the earliest pre-pottery Neolithic (12 000–10 500 Cal. yr BP) there are clear suggestions of experimental or proto-agriculture, with morphologically wild cereals found outside their natural habitats, notably in high-water-table alluvial environments such as the Euphrates floodplain at Mureybit (Sherratt, 1980). By the later pre-pottery Neolithic (10 500–9300 Cal. yr BP) fully fledged farming villages appeared widely throughout the so-called Fertile Crescent, and indeed beyond it into western Turkey. Sites were much larger, reflecting communities whose populations were numbered in hundreds or even thousands, and whose mode of production was largely agricultural.

Sites of domestication were therefore restricted to the relatively small nuclear area of the Fertile Crescent, which broadly corresponds to the modern distribution pattern of wild barley stands. It is unlikely that the wild progenitors of domestic crops have remained unchanged in their location since the beginning of the Holocene, however. Finds of the two main kinds of early wheat – emmer and einkorn – at Neolithic sites show an exactly reversed distribution pattern from that which exists among their wild relatives today, for example.

Evidence therefore shows that the change from a mobile, hunter-gatherer economy to sedentary agriculture in the Near East was not instantaneous, but involved a transitional stage of experimentation, partial dependence on domesticates, and variability in individual site economies. Only when the new systems of food procurement had proved their worth were the various elements brought together and adopted as a single economic 'package'. Some researchers (e.g. Higgs, 1972) have aimed to demonstrate that the earliest evidence of domestication for individual species in the Near East lies not in the Neolithic but in the preceding Mesolithic period. Though significant in itself, this should not cause us to forget that fully agricultural economies only emerged in later pre-pottery Neolithic times, *c*.10 000 Cal. yr BP. It was the adoption of this Neolithic agricultural package rather than its original creation that occurred quickly, typically within a few hundred years. At Abu Hureyra, for example, there was an abrupt and permanent shift from gazelle (>80 per cent of all animal bones) to sheep-goat (*c*.70 per cent) between early and late pre-pottery Neolithic levels (Legge, 1996). One of the reasons for the success of the cereal-based Neolithic economy was its incorporation of animal domesticates to form an integrated agricultural package. This package was subsequently 'exported' from the Near East into adjacent areas, especially northwestwards to form the basis of traditional European peasant farming.

Mesoamerica

The major cereal crop of New World origin is maize (*Zea mays*), whose probable place of origin is central or Mesoamerica. For many years P.C. Mangelsdorf (1974) argued that the wild ancestor of maize was no longer to be found, having been made extinct, just as wild cattle or aurochs (*Bos primigenius*) have been in Eurasia. It is now believed, however, that ancestral maize was probably descended from teosinte (*Zea mexicana*), a wild relative of maize still found in Mesoamerica (Wilkes, 1989; Martienssen, 1997). Teosinte has a small and fragile ear (cob), seeds embedded in the rachis, and hard glumes – in other words, similar features to those that differentiate wild from domestic cereals in the Near East.

Projects at Tehuacán and Oaxaca in southeast Mexico have traced the Holocene evolution of maize cultivation in bioarchaeological records. Tiny maize cobs less than 3 cm long have been recovered from archaeological contexts dated back to almost 8000 years old, and supplementary evidence such as cereal pollen suggests that even before this there existed the technical knowledge and potential cultigens for a farming economy. Significantly, and in marked contrast to the Near East, there appears to have been no rush in Mesoamerica to abandon hunting and gathering and take up farming. Garden cultivation of maize and other plant domesticates such as squash (*Cucurbita*), chili pepper (*Capsicum*) and avocado (*Persea*) were simply added to existing forms of food procurement (Pickersgill, 1989; Smith, 1997). Through selective pressures maize cobs became progressively enlarged over time (see plate 5.3), and a critical threshold may have been reached around 4300 Cal. yr BP when, with yields of 200 kg/ha, maize became more productive than any wild plant resource. Only at this point, it is suggested, was maize-based farming capable of supporting settled farming communities (Bray, 1977), and it is

Figure 5.2　The Near
Eastern centre of
domestication

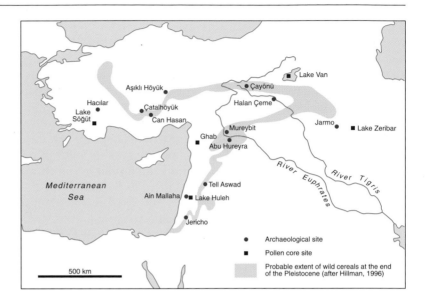

Plate 5.1　Traditional Near Eastern agriculture: this threshing sled uses flints to separate
kernels from their husks

Technical Box VI: Morphological changes in plant domesticates

Under natural conditions the seeds of cereal grasses such as wild einkorn (*Triticum monococcum* subsp. *boeoticum*, plate 5.2) would have been attached to a rachis, or central stem, which shattered when the seeds were ripe, thus aiding seed dispersal. Once harvested by sickle, on the other hand, a tough rather than a brittle rachis became advantageous. Regular harvesting by sickle therefore eventually led to the selection of mutant strains whose seed heads did not shatter when ripe. In similar fashion, threshing of the seeds tended to select strains whose grain separated from the chaff, and these became free-threshing wheats. Most of the first wave of plants to be domesticated were self-pollinating (Zohary and Hopf, 1993). This isolated them reproductively from wild progenitors and meant that farmers could grow desirable cultigens in the same area as wild varieties without the former being genetically 'swamped'.

Plate 5.2 Wild einkorn (*Triticum boeoticum*) from the Near East: one of the main ancestors of modern wheat

TECHNICAL BOX

Human pressures such as these eventually led to changes in the plant that made it a distinctively different, domestic species. Domestic wheats have more chromosomes, larger seed heads (and therefore greater yields), simultaneous ripening, and other features; but they are also dependent on human beings for propagation, and hence survival. In fact, a main diagnostic feature of domesticated plants is the loss of the ability to disperse seed without human assistance (Blumler and Byrne, 1991). Domestication can therefore only be recognized from plant morphology some time after the selective processes began, because of the time lag required for species' genetic make-up to be modified. Experimental harvesting of morphologically wild cereal stands growing has shown that this time lag will have been short (less than 200 years) even without deliberate selection, but only once certain key harvesting methods were adopted (Hillman and Davies, 1990). Harvesting of a crop by beating into baskets, for example, would leave the seed crop remaining in its morphologically wild form.

Plate 5.3
Domesticated maize cobs from archaeological excavations at Tehuacán, Mexico. The oldest one (*c.*7500 Cal. yr BP), at only 3 cm long, is hardly larger than its wild ancestor, while the most recent (500 years old) is similar to modern maize

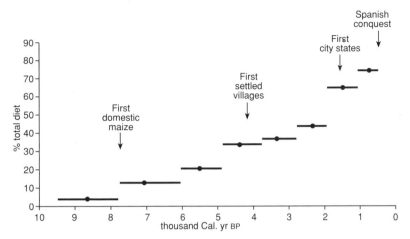

Figure 5.3 The changing contribution of domestic plants to total diet, Tehuacán, Mexico (data from MacNeish, 1967)

at this date that fully developed agriculture first appears in the New World (see figure 5.3).

A potentially different story has been suggested from the measurement of the stable isotopes of carbon (see chapter 2). Isotopic analyses of human skeletons from Tehuacán suggest that maize and other grains already formed a major component of diet as early as 7000 Cal. yr BP (Farnsworth et al., 1985). This apparent discrepancy may have been due to intensive harvesting of grasses such as *Setaria* millet, which have the same isotopic signature as maize, during the mid-Holocene. In other parts of Mexico, in any case, wild resources remained the major component of diet until much later than this (Blake et al., 1992).

Although the bio-archaeological record for Mesoamerica is less complete than that in the Near East, sufficient data exist to allow a comparison to be made between the pattern of domestication in the two regions. Similarities include a recognizable nuclear hearth area within which agriculture originated, and dates early in the Holocene for when this process began. On the other hand, the differences are no less striking. Whereas agriculture quickly replaced h-f-g as the main source of food in the Near East, in Mesoamerica this process took at least 3500 years. One reason for this delayed reaction may have been the paucity of animal domesticates in the New World. The native American llama (*Lama*) and guinea pig (*Cavia*) are relatively unimportant food resources and, moreover, they could not be integrated with crop husbandry as, say, cattle were for traction and manure in the Old World. A further difference between the two regions is that Near Eastern domestication was essentially a once-and-for-all occurrence squeezed into two or three millennia immediately after the end of the Pleistocene, but in the New World it has been a continuing process throughout the Holocene (see figure 5.4).

Figure 5.4 Recorded first appearances of individual domesticates in different regions during the Holocene

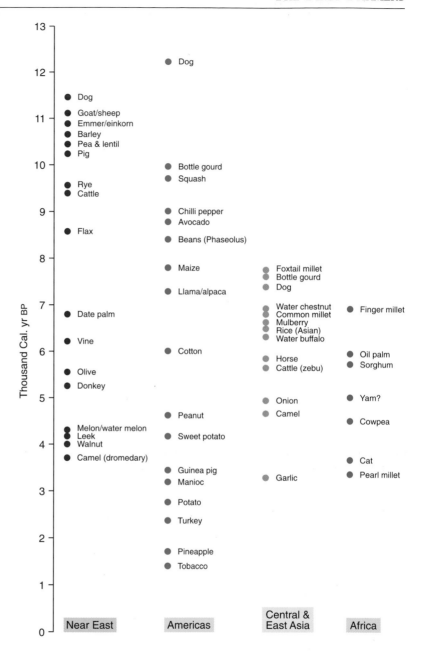

The ongoing nature of the latter has much to do with the fact that New World domestication was not restricted to Mesoamerica, even if its principal focus was there (see figure 5.1). While a few crops, such as the sunflower (*Helianthus*) originated north of Mexico, much more important have been crops derived from South America, both of highland and low-land origin. Because teosinte is not found south of Guatemala and because maize appeared in an already developed form in

South America around 6000 Cal. yr BP (Bush et al., 1989), it has often been believed that maize was exported southwards from Mesoamerica rather than being an independent South American innovation. On the other hand, the majority of maize varieties found in Peru are endemic to that country, suggesting that there may have been a secondary centre of maize domestication in South America (Wilkes, 1989). Maize apart, however, the history of New World domestication is very much one of continuing interaction and exchange of species between different regional centres. Carl Sauer characterized this in terms of a Mesoamerican origin for seed agriculture and a South American origin for vegetatively reproducing plants such as manioc. Although this distinction appears broadly valid, there is evidence in favour of multiple domestication of at least some plants; for instance beans (*Phaseolus*) appear early and independently as domestic crops in both Mexico and Peru (Harlan, 1975, ch. 12).

China and southeast Asia

The third of the world's great cereal crops is rice, of which the most important type is that indigenous to eastern and southern Asia (*Oryza sativa*). The earliest known communities with economies based on rice cultivation were those of the early Neolithic culture of China's lower Yangtse river valley, dated back to 7500 Cal. yr BP (Glover and Higham, 1996). At the site of Ho-mu-tu, waterlogged conditions led to the preservation of a layer of stalks, grains and husks of rice spread over 400 m^2 and up to half a metre thick, plausibly an ancient threshing floor. Both morphologically wild and domestic rice were found at the site, as were hoes made of ox shoulder blades, and bones of domesticated water buffalo, an animal long associated with lowland rice culture. The combination suggests that farming was based on fixed-field horticulture rather than a shifting form of cultivation. The latter may have been more appropriate in the drier loess lands to the northwest, where the contemporary Yang-shao villages depended on foxtail millet (*Setaria*) rather than on rice (An, 1989). It appears hunting and gathering had only recently been abandoned as the dominant mode of production at these Chinese early Neolithic sites, to judge from the continued importance of wild species, such as turtle and deer in animal bone remains. However, virtually nothing is known of the preceding cultures in this region. What is probable is that the origins of rice domestication are likely to be found to the south of this northern Chinese nuclear hearth area for agriculture.

The immediate ancestor of *Oryza sativa* was the annual wild rice *O. nivara*, which itself may only have evolved from the

perennial form *O. rufipogon* late in the Pleistocene. These wild
relatives of Asian rice can be found over much of southeast
Asia, even into Australia, but Chang (1976) has proposed a
somewhat more restricted area of origin for cultivated rice
stretching from northeast India through Burma, Thailand and
Vietnam to southern China (see figure 5.5). Much of this area
lay within the province of the Hoabinhian culture during the
terminal Pleistocene and earliest Holocene. A wide range of
plant and animal resources was exploited at Hoabinhian sites
such as Spirit Cave in Thailand, but there are few – if any –
signs of incipient rice cultivation here. Even at the much later
Thai sites of Ban Chiang and Non Nok Thar (*c.*6000 Cal. yr BP
onwards), carbonized rice glumes that were preserved may not
be morphologically domestic. This suggests that early manipulation of wild rice took place in the context of an intensified
utilization of a wide range of food resources in the forests of
southeast Asia during the early Holocene (Sauer, 1952). Any
cultivation was probably limited to 'dibble culture' of upland
rice and other grains (Chang, 1976). It may only have been when
rice was cultivated outside its natural habitat in the lowlands
of China that its true potential as a cultigen came to be realized. However, in the absence of detailed bio-archaeological
field research of the kind undertaken in the Near East, a detailed scheme for the origins of east Asian agriculture remains
speculative.

Tropical domesticates

The tropics have long been the 'Cinderella' regions of plant
and animal domestication. This has been explained in terms
of the poor preservation of root and tuber crops in archaeological sites, compared with seed crops such as wheat and
maize. Furthermore, asexual reproduction inhibits genetic
change and makes cultural modification of vegetatively reproducing plants harder to detect (Hawkes, 1989; Hather, 1994).
The dates at which crops such as manioc (*Manihot*), yams
(*Dioscorea*) and taro (*Xanthosoma*) were domesticated are consequently rather sketchy, although all are of undoubted tropical origin. On the other hand, the magnitude of this problem
has often been exaggerated. There are, for example, tropical
environments such as the Peruvian desert in which plant
remains are exceptionally well preserved. More significant still
is the importance of tree, fibre and seed as well as root-tuber
crops in the tropics; there are, in fact, more tropical cereal
crops than there are mid-latitude ones!

This is well illustrated in the case of sub-Saharan Africa,
where nine wild indigenous grasses are known to have been
domesticated (Shaw, 1977). One of the most important of these

is sorghum or guinea-corn (*Sorghum bicolor*), whose different domestic races exhibit clear regional patterns across the continent. From these, Jack Harlan and Ann Stemler (1976) have postulated an evolutionary sequence starting in the eastern Sudanian zone (see figure 5.6), with later secondary centres in the western Sudanian zone, east central Africa, and also in northwest India. Testing this hypothesized sequence bio-archaeologically should be no more difficult than researching the origins of wheat, but early evidence for sorghum has so far been limited to a single site – Kadero in the Sudan – dating to *c.*5700 Cal. yr BP and which produced pottery with grain impressions (Krzyzaniak, 1991). One of the few cases where early adoption of an African crop has been demonstrated is at Dhar Tichitt in the western Sahel, where an initial reliance on harvesting of wild grasses was replaced around 3200 years ago by cultivation of pearl or bulrush millet (*Pennisetum*). The paucity of bio-archaeological evidence notwithstanding, likely areas of domestication for native African crops have been proposed (see figure 5.6), and they suggest a much more spatially diffuse pattern than in Mesoamerica, China or the Near East.

Clear centres of domestication, of the type proposed by Vavilov, do not seem to be typical in low latitudes generally. Thus in South America, manioc is a crop of lowland origin – perhaps from the Orinoco basin – while the potato (*Solanum*) is native to the Andean mountains (Pickersgill and Heiser, 1977; Hawkes, 1989). However, any temptation to regard tropical agriculture as a chronologically late offshoot of mid-latitude cereal cultures should be resisted. Domestic crops appear as early as 9000 years ago in South America, while the first farmers on the south side of the Sahara grew locally domesticated crops, not imported ones. But the most dramatic – and unexpected – evidence for early independent agriculture in the tropics comes from the highlands of Papua New Guinea. Agricultural developments involving both swamp and dry-land cultivation of yams, taro, sugar cane and bananas, probably along with pig husbandry, occurred well back into the Holocene. At present, the earliest records come from excavations of field drainage systems at Kuk swamp (Golson and Hughes, 1980; Bayliss-Smith, 1996). These date back to 6800 Cal. yr BP, with geomorphological evidence of swamp disturbance as early as 10 000 Cal. yr BP and pollen indicative of forest clearance appearing by 5700 Cal. yr BP (Flenley, 1988). This potentially puts Papua New Guinea within a whisker of having the oldest known agriculture anywhere.

Unlike the Near East or Mesoamerica, on the other hand, Papua New Guinea does not seem to have acted as a major centre of diffusion for its newly acquired plant and animal domesticates. For all its proximity to Papua New Guinea, for

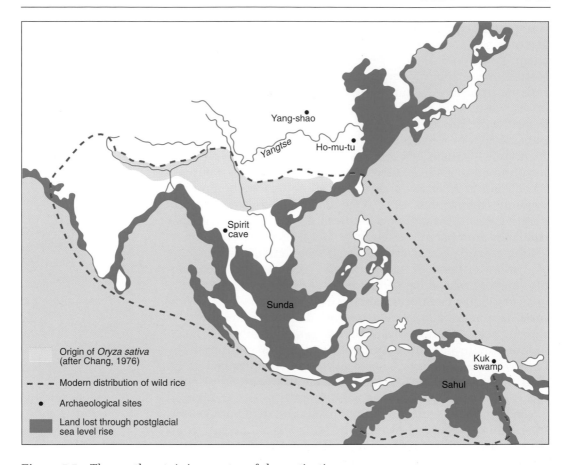

Figure 5.5 The southeast Asian centre of domestication

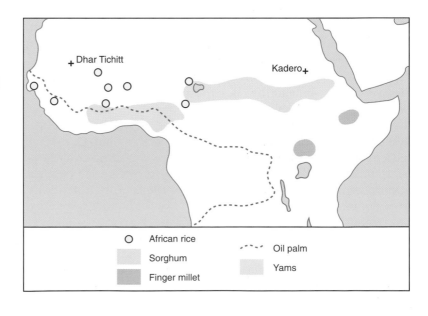

Figure 5.6 Probable
area of origin of
African plant
domesticates (data
from Shaw, 1980 and
Harlan, 1976)

example, agriculture failed to reach Australia prior to European contact – or rather, it failed to be adopted there. The fine
dividing line between farming and foraging is well illustrated
by aboriginal Australia, where women customarily exhort plants
to be generous and yield a big tuber as they dig them up. Once
out of the ground, no matter how large the tuber, tradition
decrees that the woman should now complain and berate the
plant, 'Oh you worthless plant, you lazy thing, you stingy
plant. Go back and do better'. Saying this, she would chop off
the top of the plant, put it back in the hole from which it
came, and urinate on it. This apparently ancient practice
resembles deliberate propagation so closely that we are left
wondering how the next step – that to true cultivation – could
have been avoided (Carter, 1977, p. 95, citing Tindale).

Independent innovation or diffusion?

Much early writing on plant and animal domestication was
influenced by diffusionist ideas. Agriculture was thought to be
such an extraordinary advance over hunting and gathering that
it could only have been 'invented' once, from whence it spread
to all – or at least most – corners of the Earth (Carter, 1977). As
for its place of origin, Griffith Taylor pointed to the Near East,
specifically to Egypt, while Carl Sauer believed in a southeast
Asian locus. Diffusionist models, in addition to being ethno-
centric, also had to invoke intercontinental migration of prehistoric peoples, for instance from the Old to the New World,
or at least close culture contact between them.

One domesticate sometimes used to support the model of
cultural dispersal is the bottle gourd (*Lagenaria siceraria*). This
holds pride of place among domesticated plants for its early
occurrence in areas as distant as Thailand and Mexico. One
speculation is that its wide distribution was caused by gourds
– which are buoyant – floating from one continent to another.
A more likely explanation, however, is that the gourd was a
pantropical species which, as Jack Harlan (1975, p. 239) put it,
is simply 'the kind of plant that anyone would use'. In particular, it makes an excellent container, especially for water,
which would have had obvious advantages for non-sedentary,
h-f-g populations. Compared with pottery containers, gourds
are light to carry, durable, and available free of charge from
nature. Another pantropical domesticate is the coconut (*Cocos
nucifera*). This is most likely to have an origin in southeast
Asia or the western Pacific, but it was already present on the
Pacific coast of the Americas when the Spanish arrived there
(Maloney, 1993). Like the gourd, the coconut will float and its
fruits can stay viable in salt water for over three months.

A more likely candidate for early cultural diffusion is the dog (*Canis*), which was present in China, the Near East, northwest Europe and North America by the early Holocene (Clutton-Brock, 1977). Significantly, several archaeological sites at which dog is present belonged to mobile hunter-gatherers rather than to agriculturalists. Wild dogs were camp followers of Pleistocene human groups, scavenging for animal bones after kill sites were abandoned (Binford, 1981). Given this reliance on human activities it is hardly surprising that 'man's best friend' followed us into the Americas and became our earliest domestic animal. Then, as now, dogs would have acted as camp guards, but in areas such as Mesoamerica they also provided an item of diet.

However, bio-archaeological evidence taken overall indicates that global diffusion of domesticates was rare prior to the last 500 years. Farming systems based on the three great grain crops – wheat, rice and maize – have independent centres of origin, and the list of plant and animal domesticates for different parts of the world shows virtually no overlap (see figure 5.4). American domesticates, for example, are derived from wild species indigenous to the New World, with species of African and Eurasian origin conspicuously lacking. In fact, it seems remarkable that there are not more than the handful of species, such as the dog and bottle gourd, which are common to more than one region of domestication. This is not to deny that domesticates were exchanged between adjacent regions, such as Peru and Mesoamerica, nor that farming was diffused from the Near East into Europe, but it does deny that agriculture as a global mode of production has a single point of origin. Within any particular centre of origin, however, genetic evidence suggests that domestication of primary crops like emmer wheat, chickpea and maize is most likely to have been accomplished only a few times (Zohary, 1989; Blumler, 1992).

The Role of Environmental Change in Early Agriculture

Agriculture is often seen as historically inevitable, but its advantages over an h-f-g mode of production are not as obvious as they might at first appear. For example, a harvest from dense natural stands of wild einkorn in southeast Turkey yielded a kilogram of clean grain per hour (Harlan, 1967). The grain was of comparable nutritive value to modern cultivated wheat, but involved no expenditure of labour on preparing the ground, sowing or weeding. Consequently, about 50 kcal of energy were netted for every 1 kcal expended, much more than the average of 17 kcal reported for a range of subsistence agricultural systems by Black (1971). There are numerous similar ethnobotanical examples of wild rice and *Panicum* millet gathering (Harlan, 1975, p. 15ff.; Harris, 1984). We can assume that hunters and gatherers are unlikely to have taken on the additional laborious tasks required for agriculture unless they had a strong incentive to do so. Explanations for the beginnings of agriculture have therefore often involved some shift in the balance between populations and their food resource base, either because of population increase or environmental change.

One of the best-known theories relating climatic change to domestication was proposed by Vere Gordon Childe as long ago as 1929 for the Near East. At that time it was believed that high-latitude glacial periods were accompanied by lower-latitude pluvials or wet phases. Evidence of former high water levels in lakes such as the Dead Sea was used to support this glacial–pluvial correlation. Childe logically surmised that the beginning of the Holocene should have witnessed a progressive drying out – or desiccation – of the Near East, with water resources increasingly concentrated in locally favoured habitats such as river valleys. In these few 'oases', plants, animals and human beings would have been forced to reside in close proximity on a reduced resource base, conditions that rapidly led to new relationships in the form of domestication (Childe, 1935). Childe's oasis – or propinquity – hypothesis of course fitted well with archaeological evidence for an early Holocene 'Neolithic revolution' in the Near East.

The first serious test of Childe's hypothesized climatic explanation came with the study of sediment cores from Lake Zeribar in western Iran (see plate 5.4). Pollen analysis by Willem van Zeist showed that the present oak parkland was replaced by treeless chenopod-*Artemisia* steppe before 11 500 Cal. yr BP, a flora typical of a cold dry climate, not one of moist conditions. Further pollen studies in the Near East have confirmed the Zeribar sequence (van Zeist and Bottema, 1991) and they indicate a sequence of climatic and vegetational changes precisely the reverse of that predicted by Childe's oasis hypothesis (see figure 5.7). During glacial times wild cereals were probably found only in the southern Levant and perhaps parts of North Africa (see figure 5.2), but increases in temperature and rainfall around the start of the Holocene allowed these to spread north and eastwards to their present distribution over the Fertile Crescent (Wright, 1976, 1993). This expansion of the plant and animal resources needed for domestication may have been halted, or even reversed, however, just prior to the 'Neolithic revolution', by the climatic deterioration of the Younger Dryas stadial (*c.*13 000–11 500 Cal. yr BP). A reduced availability of wild cereals and other plant foods at this time might have acted as the stimulus for new, more intensive forms of human use of these resources (Moore and Hillman, 1992).

Plate 5.4 Lake Zeribar in western Iran: cores from this site have produced a record of vegetational and climatic change from the late Pleistocene to the present day

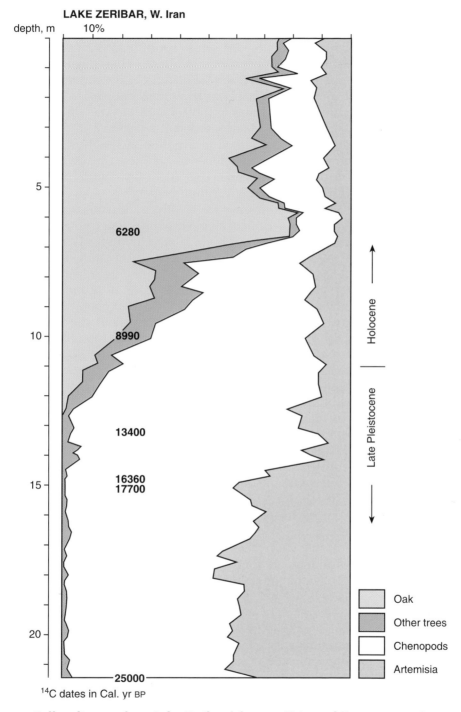

Figure 5.7 Pollen diagram from Lake Zeribar (after van Zeist and Bottema, 1977)

In Mesoamerica, an oasis hypothesis for plant domestication does not appear particularly consistent with a desiccation hypothesis either. Moisture conditions during the late Pleistocene and Holocene, inferred from pollen diagrams, lake levels, packrat midden macrofossils and other sources, are shown in figure 5.8. Mesoamerica itself does not show any clear regional climatic trend during the early Holocene. On the other hand, northwestern Mexico and southwestern United States experienced a very marked desiccation at this time, with water resources concentrated in river valleys, springs and other oases. Yet it was not here as one might have anticipated but in central Mexico that cultivation began.

These contrasting explanations illustrate two of the basic models most often invoked to link environmental change to early agriculture. Childe's oasis hypothesis is based on the deterministic idea that necessity is the mother of invention. A deterioration in the natural resource base leads to food shortage, hence the need for a new mode of production capable of higher economic output. Despite its general unattractiveness to hunters and gatherers, agriculture has the capacity to support much higher population densities than economies reliant on wild resources. And while this model does not work perfectly for the Near East or Mesoamerica, there is a better fit in some other contexts, particularly in Africa, where domestication in the Sahelian–Sudanian zone was associated with Saharan desiccation after 5500 years ago. In any case, determinist models are not always as simple as Childe's, but have also involved population as a dynamic variable (Cohen, 1977) or envisaged connections between groups in areas of better and worse resource endowment (Binford, 1968).

In contrast to this is the opportunist model, in which resources made newly available as a result of environmental change are seized upon by humans who have 'ever had an eye for the main chance'. If rising sea levels put marine food resources on your doorstep, you hunt less and fish more, as the people of Franchthi cave did at the end of the Pleistocene (see figure 4.2, p. 93). If climatic amelioration creates extensive stands of wild wheat, you learn to harvest them, as the Natufians did in Palestine around that same time. Under these circumstances, the question of whether a particular species is domesticated or not can become relatively arbitrary (Donkin, 1997). How, for instance, should we classify sheep still breeding in the wild, but loosely controlled through the provision of salt? Or wild bulls used to sire domestic female cattle? Human manipulation of plants and animals has an ancestry considerably older than farming proper, and experimentation with plant and animal domestication has to be seen in this wider context (Rindos, 1984; Harris, 1996b). Advanced hunters and gatherers – doubtless excellent botanists and zoologists – would consequently have probably been surprised at our 'nit-picking' distinctions between one form of manipulation and another. Much more important was the change from experimentation with domesticates (i.e. innovation) to their widespread adoption in fully agricultural economies.

A problem with both determinist and opportunist models is the large spatial and temporal scales over which they have been applied. Childe's oasis hypothesis, for example, was related to the whole Near Eastern region and covered a timespan of many thousand years. For Natufian or Neolithic communities, however, environmental changes over decades or centuries may have been more important than those occurring over millennia. Similarly, the important environmental resources were those immediately adjacent to their places of habitation, not those averaged over wide areas. It was for this reason that Claudio Vita-Finzi and Eric Higgs developed site-catchment analysis, a technique by which the resource potential accessible from individual sites can be assessed (Higgs and Vita-Finzi, 1972). This technique helps focus attention on site-specific resources and site-specific environmental changes. The value of the approach can be seen from a comparison of the contrasting site catchments of an early Near Eastern farming settlement – Çatalhöyük – and a late pre-farming site – Ain Mallaha.

Case study: Ain Mallaha and Çatalhöyük

Ain Mallaha is an open Natufian encampment located near a spring at the western edge of the upper Jordan valley (Perrot, 1966). It lies at the junction of several habitats, including the now-drained Lake Huleh, extensive valley marshlands and upland hills which support oak parkland with grasses including wild cereals (see figure 5.9). Only about 15 per cent of the land within a 5 km radius of the site is potentially good arable land, and most of this is not of an alluvial type. Ain Mallaha is therefore ideally located to support a broad-spectrum economy. But was the resource base significantly different at the time the site was occupied 13 000 years ago? Several sediment cores have been taken from Lake Huleh, of which the most useful in terms of vegetation history is one studied by Uri Baruch and Sytze Bottema (1991) a few km from Ain Mallaha. This core covers the period from about 20 000 years ago according to calibrated ^{14}C dating. A pollen diagram indicates that oak parkland somewhat denser than its modern counterpart covered the surrounding hills between about 15 000 and 12 000 years ago. So plant resources such as acorns and wild cereals were certainly available locally to the inhabitants of Ain Mallaha.

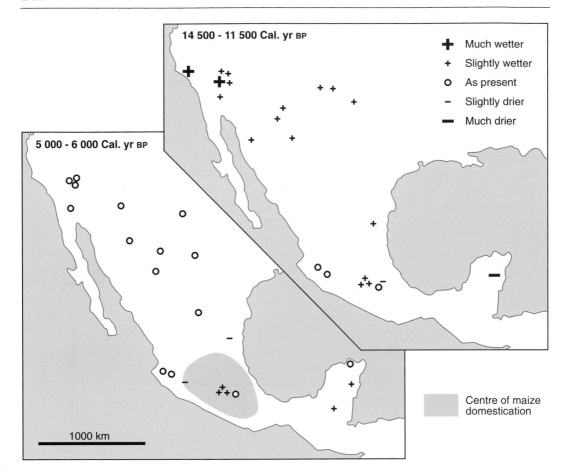

Figure 5.8 Past moisture conditions in Mexico and southwest USA (partly based on Street-Perrott et al., 1985)

Other cores from the lake also record changes in the water level of Lake Huleh (Cowgill, 1969). The uppermost sediments are peaty and indicate a shallow, eutrophic lake, but beneath this are calcareous lake sediments deposited under deeper water conditions. The existence of a larger and deeper lake around the time when Ain Mallaha was inhabited is confirmed by the presence of shoreline gravels at the archaeological site itself. The Natufian community therefore lived near the edge of Lake Huleh, which had expanded to cover what is today valley marshland. Fish, wildfowl, turtles and freshwater shellfish would have then been abundant, and it is not surprising that their exploitation is testified from excavations, for instance in the form of fish hooks and net weights. Moreover, the availability of these plant and lakeside food resources is certain, regardless of what climatic conditions were responsible for producing them. Higher tree pollen percentages and lake levels were probably a result of a slightly cooler, moister climate, but they could also be plausibly explained by other factors such as historical deforestation and damming of the lake's outlet

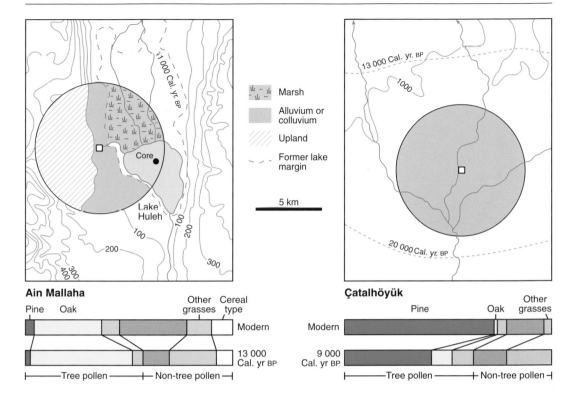

Figure 5.9
Comparative
archaeological site
catchments and pollen
records, Ain Mallaha
and Çatalhöyük

by volcanic or tectonic activity. From the perspective of Ain Mallaha's inhabitants, it hardly matters which.

The second example – Çatalhöyük – is one of the largest and most important Neolithic farming settlements known in the Near East (Mellaart, 1967; Hodder, 1997). The site lies on the flat alluvial Konya plain on the Anatolian plateau and was occupied from about 9000 until 8200 Cal. yr BP. It is located in an area of low environmental diversity, being completely surrounded by fine-grained alluvial soils (see figure 5.9). While unsuited to a broad spectrum economy such as that practised at Ain Mallaha, Çatalhöyük provided resources for crop cultivation, in particular water-retentive soils which helped augment the limited rainfall. The success of farming here is reflected in the large quantities of charred wheat, barley and other cultigens found at the site during archaeological excavations. The other main component of Neolithic agriculture in Konya was animal husbandry, in particular of cattle, which were probably locally domesticated from wild herds that grazed on the plain.

Like Huleh, the Konya basin was occupied by a lake during the late Pleistocene. In fact, Çatalhöyük lies well inside the limits of this former lake as marked by a fossil beach spit south of the site. ^{14}C dating of this and other shorelines around the basin shows that the main lake phase drew to a close about 20 000 Cal. yr BP, but that smaller residual lakes existed north and east of Çatalhöyük as late as 13 000 Cal. yr BP (Roberts, 1983). Equally important, off-site stratigraphy indicates that several metres of alluvial sediments have been deposited around the mound since 7500 yr BP by the Çarşamba river (Roberts et al., 1997a). Present-day catchment soils do not therefore provide a reliable guide as to those existing during the Neolithic. Fine-grained alluvium began to be deposited on the former lake bed at Çatalhöyük only just before the site was founded.

Evidence of vegetation history from Çatalhöyük's immediate catchment area comes from a number of pollen diagrams elsewhere in the Konya basin, along with wood charcoal and seed remains from the site itself. Together they indicate that the period 13 000–8500 Cal. yr BP was one of transition. The Taurus mountains to the south, having been largely treeless during the late Pleistocene, were re-invaded by birch, deciduous oak, juniper and pine during this time, largely in response to rising temperatures (Bottema and Woldring, 1984). In combination this range of evidence shows that the emergence of agricultural settlements on the Konya plain has to be set against a background of substantial environmental changes during the immediately preceding millennia. Between 13 000 and 9000 Cal. yr BP wetland resources diminished in the centre of the plain, while at the same time fine-grained alluvial soils were being created around the former lake margin. Climate and vegetation also came to be recognizably modern in character.

Çatalhöyük and Ain Mallaha had contrasting resources in their site catchments in the past no less than they do today. The switch from a broad spectrum to a farming economy required new resources and hence new site locations. A shift to alluvial site environments was widespread in the Near East at the beginning of the Holocene, and was coincident with the onset of Holocene alluviation in river valleys. The exploitation of these new alluvial soil-water resources represents a form of 'geological opportunism' on the part of the first farmers (Vita-Finzi, 1969a; Limbrey, 1990; Roberts, 1991).

Early Agricultural Impacts

Environmental changes associated with the last glacial-to-interglacial transition were undoubtedly instrumental in bringing about the adoption of agricultural modes of production. But the early Holocene brought opportunities to h-f-g peoples; it did not pre-determine social and economic evolution. Similar environmental opportunities had existed during earlier Pleistocene interglacials, but earlier hominids had been unable or unwilling to exploit them. An opportunist model linking environmental change and early agriculture therefore needs to be set in the broader context of post-Pleistocene cultural and ecological adaptations (Binford, 1968). This helps explain why agriculture developed independently in several parts of the world, and why it took a different course in each. Thus animal domesticates were an integral part of early agriculture in the Near East, but not elsewhere; wild food procurement was abandoned rapidly in Europe and the Near East, but not in Mesoamerica; nuclear hearth areas exist for mid-latitude cereal-based cultivation, but not for tropical agriculture; and so forth.

In any case, the advent of agriculture was not only influenced by environmental conditions, but also in turn brought tremendous potential for modifying natural environments. The most immediate impact was on the domesticates themselves. Cultural pressures did not stop with domestication, but continued to select the most productive forms, which in crops has produced a general increase in size, especially of the plant organ used by people. Many modern cultigens, such as maize, are consequently hardly recognizable as the descendants of their wild ancestors (see plate 5.3). Domesticates often also lose their natural means of dissemination and become dependent on agricultural practices for their propagation. But if cultigens could not survive without human help, neither could humankind survive without cultivated crops. They have domesticated us as much as we have domesticated them.

Agriculture thus strengthened the mutual dependence between people (as farmers) and a limited range of (domestic) plants and animals. In so doing, it also brought humans into conflict with other elements of nature. The artificial environment created by agriculture protected crops and farmyard animals from predators and advantaged them in relation to natural competitors. The ecological consequences were sometimes greater for these wild species than they were for the domesticates themselves. Sheep and cattle are numerous in modern Britain after all, while predators such as the wolf (*Canis lupus*) and competitors such as the aurochs (*Bos primigenius*) are now extinct. Extinction has occurred both directly, through hunting kill-off, and indirectly, through loss of habitat. The latter relates to the most important aspect of all in the new human relationship with nature – farming's use of the land. Cultivation of crops requires at least partial clearance of the existing vegetation cover and its replacement by field plots. Consequently, early agriculture is associated with the first substantial human impact upon the soil. This impact was all the more permanent because of agriculture's association with a settled, or sedentary, way of life.

Sedentism is known in a few non-farming societies such as the Kwakiutl Indians of Canada's Pacific coast. Fully sedentary communities probably existed around the former lakeshores of the basin of Mexico from 9000 Cal. yr BP on, over

four millennia before agricultural villages developed there
(Niederberger, 1979). However, in both cases sedentism was
only possible because of locally favourable habitats in which
resources such as wildfowl and fish were available year round.
It is similarly true that some agricultural societies, notably
those employing shifting cultivation, can exhibit varying de-
grees of mobility. None the less, it remains the case that a
sedentary way of life is much easier under a productive than
under a predatory economy. Cultivated fields require attention
through much of the agricultural year, while the need to store
harvested crops also reduces freedom of movement. Peasant
farmers are, in a real sense, tied to the land.

Some of the best-recorded examples of early agricultural
impacts come from Neolithic Europe, and these are used here
to illustrate the diffusion and environmental relations of the
first farming communities.

European agricultural dispersals

Figure 5.10 The
spread of Neolithic
farming across
western Eurasia
during the early
Holocene

Between 9000 and 5500 Cal. yr BP Neolithic farming spread
across Europe from the Near East, primarily northwestwards
along the Danube–Rhine axis (see figure 5.10). Most of this
was accomplished by two great leaps in the frontier of settlement; one from the Aegean to the Great Hungarian Plain (8200–
7800 Cal. yr BP), and a second from there across the North
European Plain (c.7400 Cal. yr BP). The lack of continuity be-

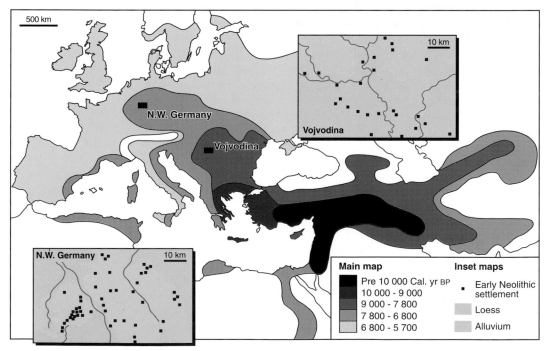

tween late Mesolithic and primary Neolithic sites strongly suggests that this spread occurred by colonization by pioneer Neolithic farmers, so-called 'demic diffusion' (van Andel and Runnels, 1995). At sites such as Franchthi cave, where there was continuity of occupation through the Mesolithic–Neolithic transition, the food-procurement system changed abruptly at this time (Hansen, 1991), suggesting that new people had arrived. Some authors have seen an analogy with the frontier of European settlement in North America during the eighteenth and nineteenth centuries AD, and have tried to measure the rate at which the population 'wave-front' advanced across Europe (Ammerman and Cavalli-Sforza, 1971). However, Neolithic farmers only exploited a small portion of the total landscape, selecting those particular habitats – notably alluvial and loessic soils – best suited to their agro-ecosystems. In the western Mediterranean and parts of northern Europe (e.g. the Ertebølle culture in Denmark), sites with transitional economies do occur, indicating that here agriculture may have been adopted by pre-existing Mesolithic populations. Environmental and cultural factors, as well as demographic ones, therefore determined the rate at which early farming spread (Dennell, 1983; Zvelebil, 1986).

The first farmers of southeast Europe operated agro-ecosystems strongly reminiscent of those developed in the Near East. Sheep/goat and emmer wheat were economic staples, but now they were living outside their natural ecological zones and could only survive with human protection. The crops on which the system depended were cultivated by intensive garden horticulture on hydromorphic alluvial soils (Sherratt, 1980). These alluvial environments were attractive to early agriculturalists because they provided a reliable supplementary source of soil moisture in a climate where rainfall was, as it still is, one of the main factors limiting plant growth. Cereals, once transferred from their natural habitats on to water-retentive soils, raised the carrying capacity of one small section of the environment out of all recognition (Jarman et al., 1982). As at Çatalhöyük, most alluvial environments in southeast Europe had only taken on their modern form early in the Holocene (van Andel and Runnels, 1995).

The natural environment placed limits on the form that Neolithic agro-ecology took, and once it reached the Great Hungarian Plain it encountered environmental conditions sufficiently different from those in the Near East that it was forced to re-orientate. The contrast in Neolithic settlement locations between one side of the Hungarian Plain and the other is very marked, with alluvial soils used on the east and loessic ones on the west (Kosse, 1979). The Neolithic Linear Pottery (or LBK) culture found in western Hungary subsequently spread with great rapidity through central and northern Europe (Keeley, 1992). LBK cultivation remained based on cereals and pulses, but animal husbandry switched from sheep and goat to a reliance on cattle and pig – the only species whose wild progenitors were available for local domestication in temperate Europe (see plate 5.5). Alluvial soils were no longer required because moisture deficiency was not a clear limiting factor for crop growth in central and northern Europe. But why loessic soils were so deliberately selected as their replacement is less clear. Loess certainly produces fertile soils with a large nutrient store and a loamy texture suitable for hoe cultivation. These soils may also have been chosen because they were only rarely occupied by pre-existing Mesolithic groups, as shown by detailed archaeological surveys, such as that on Germany's Aldenhovener Platte (see figure 5.10, inset map). Expanding Neolithic farmers may have been sucked into the empty portions of the landscape.

The first farmers of temperate Europe may have operated an extensive system of land use based on shifting – or swidden – cultivation, rather than on garden horticulture. This form of agro-ecology has until recently been practised in some of Europe's boreal forests, for example in northern Finland (Lehtonen and Huttunen, 1997), although its most widespread occurrence is in the humid tropics. Evidence for shifting cultivation by European Neolithic farmers comes from their great wooden long houses, each occupied by a single extended family and their livestock. Unlike the long-lived and permanently occupied village tells of the Balkans and the Near East, the central European long houses were only occupied for a limited number of years before being abandoned. At the LBK site of Bylany in Bohemia, a 50-year agro-ecological cycle has been suggested, with each house site occupied for 10–15 years each (Soudsky and Pavlů, 1972). This cycle was repeated many times in any one area. Initial forest clearance would have involved burning as well as tree felling, partly to increase soil nutrients. After cultivation ceased, shrub and some tree vegetation recolonized the land, allowing soil fertility to recover. Modern swidden agriculturalists in the tropics have a sophisticated knowledge of environmental resources and their potential (Richards, 1985), and the identification and manipulation of the loessic soil-vegetation ecosystem by pioneer farmers suggests that this was also true of Neolithic Europe.

Ecological consequences of early European agriculture

A swidden agro-ecological system might be expected to have had a more intensive impact on native European forests than

Mesolithic woodland management, and a wider one than the alluvial garden horticulture of southeast Europe, neither of which has left a clear mark in most pollen diagrams (Willis and Bennett, 1994). So, to what extent are vegetation clearance, cereal cultivation and woodland regeneration indicated in mid-Holocene pollen diagrams, and do they support the 50-year-long fallow cycle suggested at Bylany? In a paper published in 1941, Johannes Iversen was the first to recognize clearance or, as he called them, *landnám* phases in pollen diagrams. In some respects his *landnám* pollen phases conform well to the 'slash and burn' shifting cultivation model. They include an initial clearance stage, in which tree pollen declines relative to herb and grass pollen; a farming stage, in which grasses including cereal-type and weedy species reach a maximum; and a regeneration stage, in which shrubs such as hazel increase before declining as more substantial trees replace them. The clearance phase is also sometimes associated with a rise in the frequency of charcoal, suggesting that fire was employed in a 'slash and burn' manner.

In other respects, however, pollen data have produced some surprising results. Iversen originally estimated a duration of 50–100 years for the *landnám* cycle, but subsequent ^{14}C measurements have shown that they lasted much longer, generally between 200 and 600 years (Smith, 1981). Furthermore, the clearance cycle is rarely repeated in Neolithic pollen diagrams. Swidden agriculture may in fact have involved a short 5–10-year cycle, rather than a long 50-year one, but this is hard to identify in most pollen diagrams where the sampling interval between pollen spectra is usually more than ten years. Short fallow cycles would simply become blurred together into a general phase of vegetation disturbance (Edwards, 1979). Pollen diagrams, let alone high-resolution ones, are not always located close to Neolithic settlement sites and they are virtually absent from drier soil types such as loess. The contrast between an actual *landnám* phase, as recorded in a pollen diagram from Ballyscullion in Ireland, and that expected under long fallow shifting cultivation is shown in figure 5.11. It has even been suggested (Göransson, 1986) that there may not have been an expansion in cultivated land during the Neolithic 'farming phase', but simply better pollen dispersal in a more open landscape (Edwards, 1993).

As figure 5.11 also shows, the effect of Neolithic clearance on the overall woodland cover was rather small, tree pollen continuing to be dominant right through the *landnám* event. Indeed, Ian Simmons and John Innes (1996) suggest that the first cereal growing may have been slipped into patches of land that had been cleared for other purposes, such as attracting deer. More significant were changes that took place in the

Plate 5.5 Wild boar (*Sus scrofa*) from central Europe: the ancestor of the modern farmyard pig

composition of both woodland and open-habitat vegetation. One of the species affected was the understorey shrub hazel (*Corylus*). Hoe-based shifting cultivation does not require felling of large trees, which can be left standing inside field plots. By contrast, hazel scrub was cleared during cultivation phases, but regenerated quickly in the succeeding – and longer-lasting – fallow periods and thrived in the open, well-lit conditions. Hazel can survive the repeated firing of 'slash and burn' because it coppices freely, and the hazel rods produced from the shoots were of considerable practical value. They were, for example, the main source of wood used in the buried Neolithic trackways of the Somerset Levels of southwest England (Coles et al., 1973). Higher, but rather fluctuating, levels of hazel pollen characterize many northwest European woodlands in early and mid-Holocene times (Huntley, 1993b), and it has been suggested that this was a result of prehistoric cultivation and woodland management (Smith, 1970).

A species whose importance in European pollen diagrams changed even more markedly than that of hazel early in the Neolithic is the elm tree (*Ulmus*). The sharp and often permanent decline in elm pollen around 5800 Cal. yr BP has probably received more attention than any other feature of European vegetation history. Numerous hypotheses have been proposed

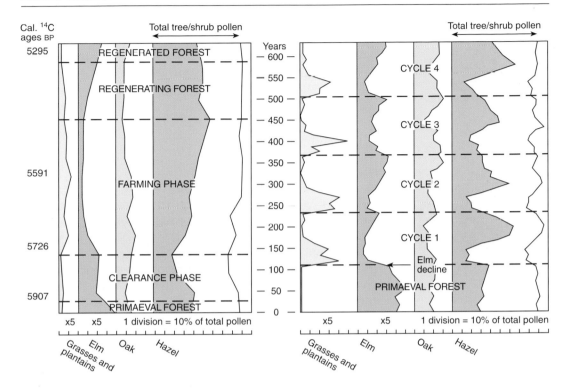

Figure 5.11 Pollen diagrams recording: (left) actual *landnám* phase at Ballyscullion, Ireland (data from Pilcher et al., 1971; Smith et al., 1971); (right) sequence expected under long fallow shifting agriculture

to explain this arboreal enigma, although they can be grouped together under four main headings (Peglar and Birks, 1993).

1 Catastrophic pathogen (disease) outbreak.
2 Anthropogenic disturbance.
3 Competitive exclusion by later migrating tree species.
4 Unfavourable climatic changes.

Although many – or all – of these could have operated in tandem, the first two hypotheses are most often invoked. The possibility of a pathogen outbreak has obviously been placed in the forefront of botanical consciousness as a result of Dutch elm disease, which has decimated Europe's elm population during the late twentieth century (Perry and Moore, 1987). The elm bark beetle (*Scolytus*), which is the main carrier of the fungus responsible for Dutch elm disease, was certainly present in Britain during the mid-Holocene (Girling and Greig, 1985). Furthermore, in a very fine-resolution pollen study of laminated sediments from Diss Mere in eastern England, Sylvia Peglar (1993) showed the elm decline to have been extremely rapid – only six years at this site – which is consistent with a catastrophic pathogen attack. On the other hand, there is a strong chronological coincidence between the elm decline and early Neolithic farming, at least in northwest Europe. As at Ballyscullion, the elm decline typically accompanies the beginning of *landnám* phases and is associated with pollen, such as cereal-type, indicative of human disturbance. Anthropogenic explanations do not generally envisage large-scale felling of elm woods to make way for Neolithic agriculture. Instead, J. Troels-Smith (1960) proposed that elm branches were lopped to provide winter fodder for stalled cattle, so reducing production of elm pollen, but not necessarily greatly reducing the number of elm trees. Pollarded trees would also have been more susceptible to attack by the disease and its carriers. So, while disease is likely to have been the immediate cause of the tree's decline, the effects may not have been so devastating or permanent had it not been for the additional impact of Neolithic farmers.

A third group of plant species to respond to Neolithic agriculture were weeds. Species such as ribwort plantain (*Plantago lanceolata*), stinging nettle (*Urtica dioica*), docks and sorrels (*Rumex* spp.) and many grasses (Gramineae) are regular appearers in post-Mesolithic pollen diagrams. These plants thrive on disturbed ground, which meant that they were pre-adapted to survive in culturally disturbed habitats, and the spread of agriculture was paralleled by the spread of weedy species (Behre, 1981). These opportunistic plants exploited human agency for their dispersal and have remained a familiar part of agro-ecosystems ever since.

One of the ironies of the European pollen record is therefore that the best indicators of early agricultural disturbance are not cultigens but the elm and the ribwort plantain! Both testify how the ecological consequences of Neolithic farming were largely indirect, even unintended. Three-dimensional pollen diagrams (Turner, 1975), such as those traversing the Somerset Levels (Coles et al., 1973), have also shown that Neolithic clearances were localized. The vast majority of Europe's wildwoods consequently remained untouched by the initial arrival of agriculture, and even those that were exploited formed part of an integrated agro-ecological system. The first farmers needed woodland, not only for timber and fodder but also – via forest fallow – to maintain the productivity of their cultivated crops. They depended on a fertile soil and clement weather for their crops, and good graze and browse for their livestock. Degradation of those environmental resources would have undermined the system of food production and threatened their own survival. Consequently, while the development of farming may have changed the relationship between people and nature, the fate of agricultural societies remained interwoven with the habitats they occupied.

CHAPTER SIX

The Taming of Nature (5000–500 Cal. yr BP)

Introduction

The middle and later part of the Holocene was the period in which the balance of power for the control of nature permanently shifted. The world some five or six thousand years ago was still primarily a product of natural forces – climate, relief and so forth. It was much as it appears in the front of atlases, with natural regions of forest and grassland, ice and ocean. Where it differed from today's 'natural regions' – for example, the absence of an arid Sahara – this was because of subsequent climatic changes, not because of human-induced ones. During the early Holocene, humans can reasonably be considered to have formed a part of the natural world, or, to take an alternative view, were dominated by nature (Simmons, 1993, 1996). This relationship derives from the fact that the majority of the world's human population were still reliant on hunting-fishing-gathering (h-f-g), a mode of production which normally uses and manipulates natural ecosystems without transforming them. Peasant farming societies, it is true, had begun to emerge, but these were as yet restricted to small parts of Europe, the Near East and southeast Asia. The long process of clearing natural vegetation to make way for cultural landscapes was only just beginning.

Contrast this situation with that at the end of the time period covered in this chapter. By the start of the modern era, around 500 years ago, many natural ecosystems had been replaced by agricultural ones, and even those retaining the vestiges of their original appearance had been altered in their detailed composition. Out of primary nature, human agency had fashioned what Clarence Glacken (1967) has termed a 'second nature'. This taming of nature was the work of agricultural societies which developed and spread over large parts of all the major land masses except Australia and Antarctica (see figure 6.1). The spread and development of agricultural modes of production created such cultural landscapes as the olive groves of the Mediterranean and the irrigated rice fields of China's river valleys.

Landscapes were not transformed by human agency alone, however. The natural environment responded dynamically to human influence, sometimes metamorphosing in a way quite different from that intended by the initial human actions. On the Pleistocene sands of northern Europe, for example, forest clearance led to soil podsolization and the creation of heathlands rather than reversion to secondary woodland (Gimingham and de Smidt, 1983). Furthermore, nature was itself actively changing independent of human action. Natural agencies such as fire, disease, sea-level change and soil maturation all continued to operate, and the climate was often far

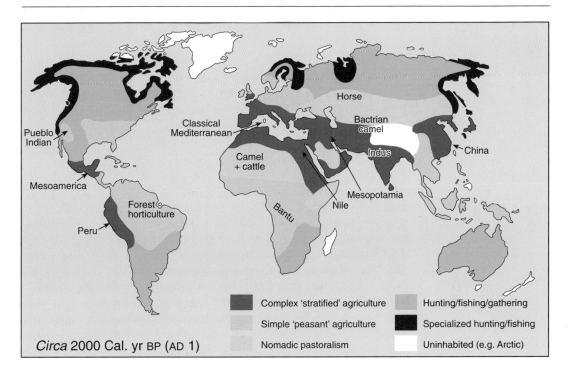

Camel + cattle
Classical Mediterranean
Pueblo Indian
Mesoamerica
Forest horticulture
Peru
Bantu
Horse
Bactrian camel
Indus
China
Mesopotamia
Nile

Complex 'stratified' agriculture

Simple 'peasant' agriculture

Nomadic pastoralism

Hunting/fishing/gathering

Specialized hunting/fishing

Uninhabited (e.g. Arctic)

Circa 2000 Cal. yr BP (AD 1)

Figure 6.1 Global cultural developments during the later Holocene (partly based on Lewthwaite and Sherratt, 1980)

from stable. Environmental change associated with these different natural factors is the first theme that will be explored in this chapter.

During the late Pleistocene and early Holocene, environmental changes had been dominated by a single forcing factor, namely, the transition from a glacial to an interglacial climate. But this unifying link was lost as climatically induced changes came to be joined and overtaken by culturally induced ones. This was linked to a progressive increase in human impact upon natural environments, and the consequent need for societies to adapt in their turn to a world which they had helped to transform. In areas such as the Pacific and the North Atlantic islands, human impact was linked to the continued geographical expansion of populations into previously uninhabited 'virgin' landscapes. Cultural changes did not have a uniform effect over the globe, however, because the pace and direction of cultural evolution varied between regions (Simmons, 1996; Diamond, 1997). It is consequently at the regional scale that human–environment relations are most usefully examined during most of the mid-to-late Holocene. For this reason, much of this chapter is devoted to selected regional environmental histories: for example in Mesoamerica, the Mediterranean basin and the British Isles. We should note that in some – although not all – parts of the world, written histories emerge after 5000 Cal. yr BP, so that historically derived chronologies become available to complement radiometric and incremental ones (see chapter 2).

Changes in the Natural Environment

Climate and vegetation

Although much less marked than during the early Holocene, the later Holocene none the less experienced some significant shifts in climate. Notable among these was the progressive cooling following the Holocene thermal optimum of around 9000–5500 Cal. yr BP. This deterioration in fact took place in a series of steps rather than gradually. One of these steps was from the warmer and drier Sub-Boreal to the cooler and wetter Sub-Atlantic period, a change dated in European peat bogs to *c.*2600 Cal. yr BP. The effect of late Holocene cooling on vegetation belts in Europe is often disguised by increasing human disturbance. The northward spread of beech (*Fagus*), for example, may have been partly due to climatic cooling which disadvantaged competing thermophilous species such as the lime tree (*Tilia cordata*). But beech was also aided by Bronze and Iron Age clearance, which provided gaps in the forest and allowed it to invade and become a dominant tree over large parts of northwest Europe (Huntley and Birks, 1983, p. 191ff.; Huntley et al., 1989; Küster, 1997). In North America, where the pace of cultural evolution was slower, there is less ambiguity about the relative effects of climatic and human agencies. The westward shift in the prairie–forest boundary and the southward movement of the boreal forest after *c.*4500 Cal. yr BP both reflect the response of vegetation to a cooler, moister climate (Webb et al., 1993).

An equally important climatic change occurred in the subtropics of the northern hemisphere, where climatic desiccation between 6500 and 4500 Cal. yr BP created the modern Saharan, Arabian and Thar deserts. Pollen records such as at Oyo (Ritchie et al., 1985) show a relatively gradual transition from savanna to semi-arid scrub to barren desert over this time period. In contrast, the water levels of some low-latitude lakes fell abruptly, suggesting that the monsoonal circulation system which controls tropical rainfall was suppressed much more rapidly (Street-Perrott et al., 1991). The change from savanna to desert not surprisingly completely altered the human ecology of the Saharo-Arabian zone. Without lakes or rivers, the 'Aqualithic' culture of the early Holocene disappeared (see chapter 4), while in the few remaining oases, such as the Nile valley, new intensified forms of land use emerged (Hassan,

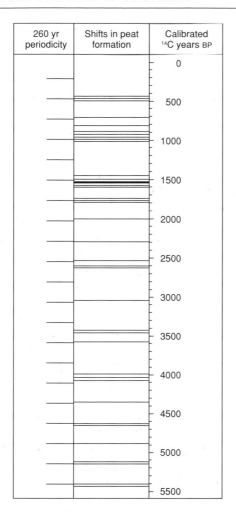

Figure 6.2 Cyclical sequence of climatically controlled recurrence surfaces in five Danish raised bogs (after Aaby, 1976)

1986). The deceleration of eustatic sea-level rise had already reduced the Nile's gradient and led to silt accumulation and the creation of the fertile delta plain (Stanley and Warne, 1993). As the level of the Nile flood declined between 5000 and 4000 Cal. yr BP (Hassan, 1988; Butzer, 1984), so simple floodwater farming became artificially regulated. New water-lifting technology such as the *shaduf* was progressively introduced, and this had major consequences for social and political conditions in Egypt (Butzer, 1976; Hassan, 1993; see also below).

Superimposed on the major climatic trends were shorter-term cycles produced by secular oscillations from wet to dry, or cold to warm. In an analysis of recurrence surfaces in Danish raised bogs, Bent Aaby (1976) found that oscillations between cool-wet and warm-dry conditions took place every 260 or, more infrequently, every 520 years since the mid-Holocene (see figure 6.2). An 800-year periodicity was found to be dominant in raised peat bog record from northern England (Barber et al., 1994). Similar century-scale periodicities have been recorded from Holocene histories of glacier advance and retreat (Grove, 1979, 1988). Whether these oscillations represent true, regular cycles is uncertain, as is the question of the causes of these climatic changes. Volcanic eruptions with large dust emissions, sunspot cycles and internally generated stochastic variations are amongst the hypotheses proposed (Lamb, 1977; Nesje and Johannessen, 1992).

Volcanic aerosols can play an important role in disturbing the Earth's climate, for example by creating a dust veil which reduces the transparency of the atmosphere. After major historically recorded eruptions, such as those of Tambora in Indonesia (1815) and Laki in Iceland (1783), contemporary writers recorded severe late frosts and peculiar hazes in regions far removed from the volcanoes themselves (Grattan and Brayshay, 1995) (plate 6.1). Benjamin Franklin, who was American ambassador in Paris during the 1780s, described a 'constant dry fog on which the rays of the sun seemed to have little effect'. Some earlier Holocene eruptions may have been substantially larger than Laki and Tambora (table 6.1), and it has been suggested that they may have had a proportionately bigger climatic impact. For example, in about 1628 BC (3578 Cal. yr BP) the Aegean island of Santorini (Thera) exploded catastrophically, forcing its Bronze Age population to flee or perish. Dendroclimatological studies have shown that the tree-rings of Irish bog oaks were unusually narrow at this time (Baillie and Munro, 1988), while bristlecone pine growth rings of the same age in California show signs of frost damage (LaMarche and

Table 6.1 Some major volcanic eruptions during the last 7000 years, identified in the Greenland (GISP2) ice-core record

Date	Volcanic eruption	SO₄ anomaly (ppb)	Volcanic Explosivity Index (VEI)
AD 1815	Tambora, Indonesia	94	7
AD 1783	Laki, Iceland	134	4
AD 1175 (775 Cal. yr BP)	Krafka, Iceland	148	4?
AD 79 (1871 Cal. yr BP)	Vesuvius, Italy	95	5
2004 Cal. yr BP	Sheveluch, Kamchatka	291	–
3578 Cal. yr BP	Santorini, Greece	145	6
4260 Cal. yr BP	Hekla (H-4), Iceland	80	–
5150 Cal. yr BP	Akutan, Alaska	175	–
5400 Cal. yr BP	Towada, Japan	174	–
6845 Cal. yr BP	Mazama, USA	141	–

Source: Zielinski et al. (1994)

Hirschboek, 1984). These have been interpreted as the result of a cooler and/or wetter climate. However, in Turkey, which is warm and dry in summer, the same climatic change gave rise to more, not less, favourable growing conditions and to broader tree-rings (Kuniholm et al., 1996). While the evidence linking tree-rings, climate and volcanic eruptions is impressive, it also suggests that in most cases the climate perturbation was short-lived, typically a decade or less.

Many secular climatic variations, in any case, do not appear to have been synchronous between different parts of the world, and their recognition in palaeoenvironmental records is complicated by the fact that other natural agencies operated over similar timescales (see figure 1.1, p. 5). Fire, for example, has

Plate 6.1 The 1970 eruption of Hekla in Iceland. Previous eruptions left tephra marker layers across northern Europe and had major ecological impacts on Iceland itself

affected seasonally dry woodland ecosystems which have a recovery time following disturbance and secondary succession of 50–200 years (Wright and Heinselman, 1973; Swain, 1973; Clark et al., 1989). Fire affects the plant and animal community as a whole, but individual taxa react differently, some being harmed while others benefit.

Another important agency in ecological change is disease pathogens, which, unlike fire, are host-specific. During the twentieth century, diseases have been responsible for the decline of several tree taxa, including the elm and the chestnut (*Castanea*). Chestnut blight was introduced into North America in 1904, and within 40 years it had killed most of the mature trees in New England and the Appalachian mountains (Patterson and Backman, 1988). Dutch elm disease has been implicated in the mid-Holocene European elm decline (Girling and Greig, 1985; see chapter 5), but it is difficult to prove a definitive link between this – or any other – former species decline and a disease outbreak. Probably the most plausible example is the reduction in hemlock (the conifer *Tsuga canadensis*, not the flower Socrates used to poison himself!), which took place throughout eastern North America at about 5500 Cal. yr BP. Fine-interval pollen analysis of laminated sediments from Pout Pond in New Hampshire indicates that the decrease occurred within only seven or eight years (figure 6.3C). This catastrophic decline appears to have been synchronous over the whole geographical and climatic range of eastern hemlock within the limits of ^{14}C dating. It was also followed by an increased concentration in the pollen of successional trees, such as birch (Allison et al., 1986). Hemlock subsequently remained rare for the following 1500 years, but eventually recovered, albeit much more gradually than its earlier precipitate decline (figure 6.3A and B).

The origin and development of blanket mires

A problem already encountered with mammalian extinction (see chapter 3) and the elm decline (see chapter 5) is that of ascertaining whether responsibility for specific environmental changes lay with human or natural factors. This problem becomes particularly acute during the mid-to-late Holocene, and a good example is the question of the origin of extensive areas of peat which today blanket the oceanic upland regions of northwest Europe. On gently sloping plateaux this peat is often raised above the general water table and receives mineral nutrients only from the atmosphere, mainly in rainfall. These **ombrotrophic** peat bogs were not present ten or even eight thousand years ago, but have developed during the mid-to-late

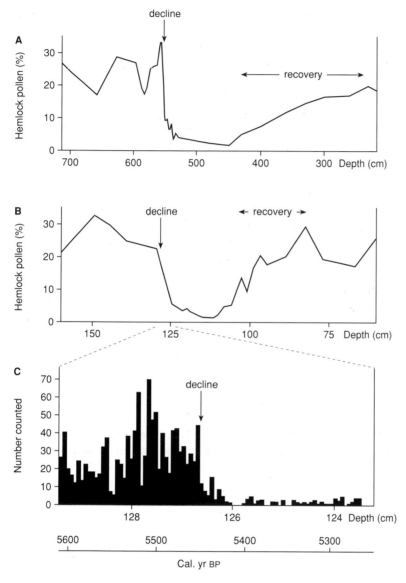

Figure 6.3 **A** and **B** Catastrophic mid-Holocene decline and gradual recovery of hemlock trees in eastern North America, recorded in pollen records from two lakes (after Allison et al., 1986). **C** Micro-stratigraphic pollen analysis from one of the lakes (Pout Pond) covering the 350 years of the initial decline

Holocene as a result of some combination of climatic deterioration, soil retrogression and human impact. Blanket peats have buried many upland soils and enabled new vegetation communities to become permanently established. For example, the blanket bog covering the high plateaux of Dartmoor, southwest England, will not revert to woodland even though relics such as Wistman's Wood show that tree growth is still climatically possible here. The key factor in peat formation is waterlogging of the soil (see figure 6.4). Waterlogging causes a reduction in cations and an increase in hydrogen ion availability (= fall in pH) which, with a suppression of microbial activity, fundamentally alters soil conditions. Only acid-tolerant plants such

Figure 6.4 Model for upland peat initiation (after Moore et al., 1984)

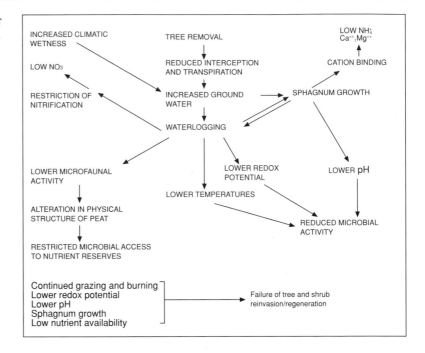

as ling heather (*Calluna vulgaris*) and cotton grass (*Eriophorum*) can survive in this environment, but the dominant component is usually different species of bog moss (*Sphagnum*). Like a sponge, *Sphagnum* is able to absorb water and hold it in the plant cells, and so maintains its own water supply. As a result the peat bog is always saturated with water.

The age and origin of blanket bog and other mires can be established by studies of peat stratigraphy, radiocarbon dating, pollen and plant macrofossil analyses. In some areas such as northwest Scotland, upland peat bogs can be regarded as climatically controlled ecosystems. Not only is there high effective rainfall here, but basal ^{14}C dates for peat initiation generally fall between 9000 and 6000 Cal. yr BP (Moore et al., 1984), that is, before the arrival of Neolithic agriculture. Elsewhere, on the other hand, many upland peats began to form at and after the time of the elm decline (Moore, 1975; Clymo, 1991). There is a strong suggestion that removal of tree cover triggered a sequence of environmental changes that led to woodland being permanently replaced by peatland (see plate 6.2). Trees, where they exist, fulfil a critical hydrological role by evapotranspiring at much higher rates than non-arboreal vegetation, and by improving drainage via their rooting system. They also help maintain the circulation of nutrients. Trees can be killed not only by felling, but also by increased groundwater, and in ombrotrophic mires groundwater levels are primarily

Plate 6.2 Tree stumps in peat cuttings testify to the existence of ancient forests on what is now blanket bog at Fallahogy, Ireland

related to the prevailing climatic conditions. Because the Atlantic/Sub-Boreal transition *c.*5700 Cal. yr BP did not mark a shift to a wetter climate, a climatic trigger for blanket bog initiation at this time seems less likely than an anthropogenic one.

Once in existence, blanket mires evolved to a stable equilibrium state which none the less included periods of slower and more rapid peat growth (Blackford, 1993) (see figure 6.2). In northern Scotland, tree stumps buried in blanket peat show that Scots pine was able to spread beyond its modern range for several centuries around 4500 years ago, probably because a period of warm climatic conditions dried out the peat surface (Gear and Huntley, 1991). Later, at the Sub-Boreal/Sub-Atlantic transition, between 3000 and 2000 Cal. yr BP, there was a pronounced change in mire stratigraphy across northwest Europe indicating the opposite trend (Barber, 1993). Dark, oxidized peat typical of slow-growing mires and often including buried tree stumps was replaced by relatively undecomposed *Sphagnum* peat typical of wetter, fast-growing mires. This humification feature is often known as the **Grenzhorizont** (boundary horizon) after the terminology created by the German botanist, Weber.

Coasts and rivers

For most of the world's coastlines, the early Holocene had been dominated by the continuing eustatic rise in sea levels as northern ice caps finally melted away (see chapter 4). Around 7000 Cal. yr BP coastlines took on their familiar modern form, and with the exception of coasts experiencing glacio-isostatic rebound, the second half of the Holocene was a period of relative sea-level stability (see figure 4.1B, p. 89). In detail, different parts of the world have each had their own local sea-level histories over the last 6000 years. In much of the Pacific, for example, sea levels lay several metres above their modern position during the mid-Holocene (Nunn, 1990), whereas on the eastern seaboard of the United States they have continued to rise gradually to their present elevations.

With sea-level stability came the formation of modern coastal landforms, such as Chesil beach in southern England and coral reef atolls in the Pacific. Above all, in estuaries and deltas, rivers started to reverse the long-lived marine incursion onto the land. As rivers deposited their sediment load so new land was built up and shorelines were pushed seawards again – a process that has continued up to the present day. In the lower valley of the River Rhône, the sea reached its furthest landward position around 7000 years ago and has been in retreat ever since, as sedimentation has created new alluvial habitats including the Camargues (Pons et al., 1979). At the mouth of the Mississippi no less than seven lobes-foot deltas have been built up and destroyed over the last 4600 years, while the Hwang-He river in northern China has pushed the coastline out some 250 km over the same period. A smaller-scale but historically interesting sequence of early Holocene marine incursion and later Holocene progradation occurred in the Scamander river valley of northwest Turkey (Kraft et al., 1980).

Boreholes show a wedge of sediments containing marine fossils sandwiched between fluvial strata, ^{14}C dated to between *c*.11 000 and 2000 Cal. yr BP (see figure 6.5). The Scamander valley was thus an embayment of the sea, and not an alluvial plain, at the time when the famous Bronze Age site of Troy (Hisarlik) was occupied. Homer's description of the landing of the Mycenaean Greek fleets during the Trojan War thus rests intact.

Inland, river systems were by now fully adjusted to regional climates and to the slopes, geology and natural vegetation cover in their catchments. Rivers, being dynamic systems, periodically abandon their channels and this led to the creation of meander cutoffs where standing- rather than flowing-water conditions prevailed. These oxbow lakes underwent hydroseral succession and many have become infilled with sediment; initially sands and gravels, later peats. The sediment archives of these old meander cutoffs are often rich in pollen and the remains of insects and microcrustaceans (e.g. cladocera), which can provide information on changing floodplain ecology (Amoros and van Urk, 1989). From this we know, for example, that most lowland English river floodplains were covered by dense alder woodland during the mid-Holocene (Brown, 1988). The dimensions of palaeochannels and the size of the largest sediment particles at the base of their infill can be used to calculate their former river discharge (Rotnicki, 1991; Bishop and Godley, 1994). Using maximum particle size data from the headwater streams of the Mississippi headwaters, Jim Knox (1993) has shown that relatively modest changes in the Holocene climate of the US Mid-West led to large increases in flood magnitude. Because big floods only happen rarely, fluvial palaeohydrology provides one of the few sources of information on long-term flood frequencies and magnitudes (Kochel and Baker, 1982).

Cultural Evolution

From a common starting point in peasant farming, there evolved during the second half of the Holocene a diverse range of cultures based on agriculture or pastoralism. The most important of these were socially stratified and involved asiatic, classical or feudal modes of production (Russell, 1989). All of these shared in common the production of an economic surplus by agricultural labour that was then taken and used by another social class. This new form of social relations also changed human–environment relations in several important ways. It meant, for example, that maximum economic output

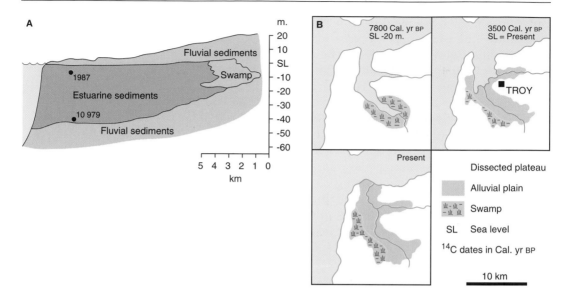

Figure 6.5
Geomorphic
reconstructions in
the vicinity of Troy
during the Holocene
(after Kraft et al.,
1980)

tended to replace the mix of social and economic goals that
characterizes communal peasant farming societies. Agricultural
output was maximized by improving the productivity of existing farmland or by increasing the cultivated area. The latter
led to clearance of woodland and draining of marshland, but
also included cultivation of marginal land susceptible to soil
erosion and other forms of degradation. The advent of complex agricultural societies distanced and often weakened the
link between people and nature. Nature became less the 'habitat' for the farmer than a set of economic resources to be managed and manipulated by the controlling group (Simmons,
1993). This was particularly true of cultures where the dominant class was urban-based, as in Graeco-Roman antiquity.
Although these complex agricultural societies shared certain
key characteristics in common, such as urbanism, they each
had their own distinctive set of environmental relations. This
is well illustrated by the asiatic mode of production, or, as
Karl Wittfogel (1956, 1957) termed it, 'hydraulic civilization'.

Hydraulic civilization in Mesopotamia

The ancient civilizations of Egypt, Mesopotamia, India and
China all developed in a specific physical context – major
alluvial river valleys under climates of low or unreliable rainfall. Karl Wittfogel suggested that the agricultural potential of
these fertile lands could only be realized by large-scale manipulation of their soil and water resources. This encouraged
the development of hydraulic irrigation schemes, constructed

under a centralized organization. The agrarian economy in these great irrigated river valleys became the function of the state, which was also the sole owner of land. The timing and distribution of irrigation water, the maintenance of canals and the collection and storage of surplus food were the responsibility of a professional bureaucratic ruling class which held total, despotic power. The combination of organized human labour and a productive environment created a distinctive and potent mode of production which transformed nature so completely that almost nothing now remains of the original alluvial landscape.

The environmental relations of hydraulic civilization can be examined by analysing the transformation of one of the world's great river valleys, that of the Tigris–Euphrates. The Tigris and the Euphrates rise to the north of lowland Mesopotamia in the mountains of Turkey and Iran. They receive winter rains which combine with spring snow-melt to produce maximum river discharge in April and May. The winter crop is planted when the twin rivers are at their lowest levels, and the rivers then rise towards harvest time, 'threatening with inundation the crops they were with difficulty persuaded to germinate' (Ionnides, 1937, pp. 4–5). Irrigation agriculture was dependent on a 'good' flood: too high and settlements and grain stores would be ruined, too low and there would be a poor crop, food shortage and famine. In addition there was the threat of river channels changing course. This occurred periodically as alluviation raised the height of distributaries, but could also receive human help when diversion canals were cut. Most of the water carried by the Tigris and Euphrates in fact never reaches the sea, but is lost as evapotranspiration from the flat alluvial marshlands. This brought another hazard to Mesopotamia, that of salinization. As water is evapotranspired, so river-water solutes are left behind and eventually become concentrated to produce saline soils. These reduce crop yields and prevent cultivation altogether above salt levels of $c.1$ per cent.

Simple floodwater farming involving use of residual soil moisture had been part of Neolithic agriculture in the Near East from the very beginning (see chapter 5, and Sherratt, 1980). It is not surprising that the first archaeological evidence of irrigation canals appears as early as 8200 Cal. yr BP, at Choga Mami on the slopes above the Mesopotamian floodplain (Oates and Oates, 1976). Occupation of the alluvial plain proper came in the protohistoric Ubaid and Uruk periods (7300–5700 Cal. yr BP) and lay along smaller distributaries of the two main rivers. These lay above the surrounding alluvial flood basins and it was consequently possible to cut through the river levées to provide irrigation water for crops planted on the basin flanks.

Table 6.2
Salinization and
changing cropping
patterns in southern
Mesopotamia

	Cal. yr BP	wheat (%)	barley (%)	ratio	barley yield (l/ha)
Protohistoric	6000	46.5	53.5	1:1.5	–
Dynastic	4350	16.3	83.7	1:5	2537
	4250	3.0	97.0	1:32	–
	4050	1.9	98.1	1:53	1460
	3950	0.0	100.0	–	–
	3650	0.0	100.0	–	897
Early Islamic	1200	9.1	90.9	1:10	–

Source: Jacobsen (1982)

The natural channels were later straightened or replaced by
completely artificial canals to build up a true irrigation network.
Historically, Mesopotamian civilization emerged into a dynastic
phase around 5000 years ago (3000 BC), with monumental architecture, urban life and the world's earliest written records.

Despite its successes, irrigation agriculture in southern
Mesopotamia encountered increasing ecological problems.
Salinization was kept at bay by alternating years of cultivation
and weed fallow in which the leguminous perennials shok
(*Prosopis farcta*) and agul (*Alhagi maurorum*) lowered the level
of saline groundwater. This did not provide a permanent solution, and carbonized grains and textual sources show a switch
between 4350 and 3650 Cal. yr BP from wheat to more salt-
tolerant barley (Jacobsen, 1982). This was accompanied by a
decline in barley yields and frequent mention in Sumerian
texts of land abandonment due to salinization (see table 6.2).
The agro-ecological crisis was instrumental in bringing about
a switch in the centre of power from southern to northern
Mesopotamia, where saline soils were less prevalent. Irrigation in northern Mesopotamia needed to be larger and more
complex, and the higher rate of siltation here meant that canals had to be regularly cleaned and maintained. On the Diyala
floodplain north of Baghdad over 10 m of alluvium has accumulated since irrigation began over 6000 years ago. Artificial
regulation at nodal distribution points and control of the source
of irrigation water led to a codified system of water regulation
and centralized organization, much as Wittfogel's model suggests. When this administration operated efficiently, for example during the Sassanian and Abbasid periods (1750–1100 Cal.
yr BP), agriculture flourished and population densities were
high. But once central power was weakened, such as after the
Mongol invasions (c.750 Cal. yr BP), there was land abandonment and disastrous population decline (see figure 6.6).
Archaeological surveys, notably in the Diyala region, clearly

LOWLAND MESOPOTAMIA

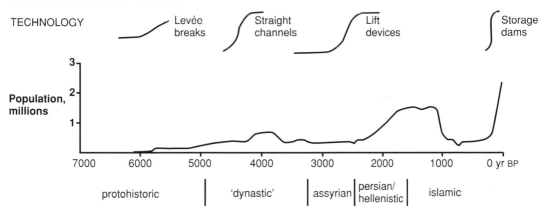

Figure 6.6 Historical changes in population and irrigation technology in lowland Mesopotamia (based on Bowden et al., 1981)

show this cyclical alternation between periods of stable irrigation and periods of disruption (Adams, 1965).

In Mesopotamia there is a strong case for a direct causal link between hydraulic agriculture and the development of a centralized political state, but irrigation systems elsewhere in the world show this link much less clearly. In Egypt, for instance, nilotic agriculture was based on an essentially natural system of flood-basin irrigation until only 150 years ago (Butzer, 1976). Unlike Mesopotamia, flood control and irrigation were locally administered and there was little need for a centralized bureaucratic apparatus. Competition for water between users was never more than a local issue in the Nile valley and water rights did not have to be rigidly codified. It is also important to recognize that while the environment may have stimulated economic and cultural developments, it did not determine them. No great hydraulic civilizations developed on the floodplains of the Murray–Darling or Mississippi rivers. Equally, both natural forces, such as soil salinization, and human ones, such as invasion from outside, could cause the complex 'machine' that was hydraulic irrigation to break down and collapse.

Environmental impact in Prehispanic Mesoamerica

Complex, stratified societies took a little longer to develop in the New World than they did in the Old, but after 3000 Cal. yr BP important indigenous civilizations did emerge in several centres within the Americas. The most notable of these were in Mesoamerica, based on maize cultivation. One of the first great cultures to emerge with ceremonial centres and monumental architecture was that of the Maya, centred in the lowlands of the Yucatan peninsula. The fact that ruined sites such

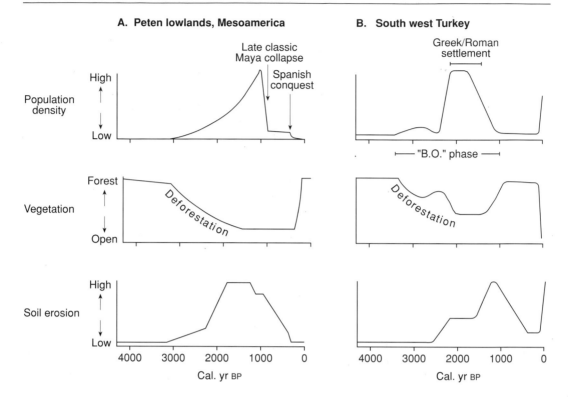

A. Peten lowlands, Mesoamerica

B. South west Turkey

Figure 6.7 Late Holocene forest clearance and recovery, linked to population and erosion, **A** Maya Lowlands (based on Binford et al., 1987) and **B** southwest Turkey

as Tikal were until recently cloaked in dense tropical forest indicates the scale of Maya landscape clearance – it was more extensive than anything that has been achieved since. The dense rural population was supported by a system of double cropping and in some areas used drained-field agriculture to exploit wetlands (Darch, 1988; Furley et al., 1995). Population levels rose progressively from 1700 to 1100 Cal. yr BP, but this brought its toll on the predominantly limestone soils, which were subject to erosion and nutrient loss. Pollen records from lakes in the Peten region show widespread late Holocene deforestation (Deevey et al., 1979; Leyden, 1987; Islebe et al., 1996), and this was accompanied by a change in the composition of the lake sediments. Forest removal and urban construction were matched by increased accumulation rates of fine-grained mineral matter to create the 'Maya clay' found in many lakes (Binford et al., 1987). Phosphorus was removed along with the topsoils, and phosphorus accumulation in the lakes closely tracks reconstructed population levels in the lake catchments (Brenner, 1983) (figure 6.7A). The Maya's increasingly precarious ecological support system appears to have been given a final push by a period of severe drought, indicated by increased salinity levels in lakes between 1200 and

1000 Cal. yr BP (Hodell et al., 1995). This environmental crisis, partly self-inflicted, and doubtless associated with political instability and social unrest, led to the sudden collapse of the Classic Maya civilization. The last long count on a Maya jade ornament dates to AD 909; the peak of aridity in a nearby lake core lies at AD 922 (Hodell et al., 1995). The ceremonial pyramids and other monuments were abandoned to the encroaching forest for a thousand years.

On the highlands of Mexico, drought was even more of a threat, and large-scale agriculture was only possible by the development of complex hydraulic irrigation schemes. The Classic period for Mesoamerican civilization at sites like Teotihuacán was similar to that in the Mayan lowlands. It, too, went into decline after c.1100 Cal. yr BP, although the collapse was less sudden and less permanent. What is sure is that Prehispanic human impact was responsible for widespread landscape degradation. Sediment records from the basin of Mexico and others in the volcanic highlands to the west record increased influx of eroded soil into lake basins, although they display regional differences in its timing as different centres went through cycles of military success and defeat (Metcalfe et al., 1989). In Lake Pátzcuaro, quantitative estimates indicate three periods of accelerated soil removal when erosion rates were at least as high as after the Spanish Conquest, and this despite the lack of the plough to turn the soil (O'Hara et al., 1993). When Hernán Cortés crossed from the coast to enter the Aztec capital of Tenochtitlán in AD 1521, the Spaniard was encountering a far from pristine land.

Pastoral nomadism

Complex agricultural societies such as existed in the Near East and Central America were not the only form of cultural development to emerge from communal peasant farming during the mid-Holocene. One of the most distinctive cultural developments was nomadic pastoralism, an extensive form of production adapted to arid or rugged terrain. Like hydraulic irrigation, pastoralism involves a distinctive set of ecological relationships between human groups and their environments, in this case mediated not through water use but through domestic animals. The domestication of animals such as cattle and sheep-goat had formed part of the early Holocene Neolithic revolution in the Near East, but a separate mode of production based on animal herding and involving a nomadic lifestyle only emerged during mid-Holocene times (Sherratt, 1981; Khazanov, 1984).

The domestication of the horse (*Equus caballus*) led to the emergence of pastoralist groups on the steppes of central Asia after 4000 years ago (see figure 3.8, p. 85), who were to be the forerunners of the later Huns, Turks and Mongols (Zvelebil, 1980). Nomadism developed around the same time in the newly desiccated deserts of Arabia and the Sahara, where the camel (*Camelus*) was found to be better suited to the arid environment (Bulliet, 1975; Köhler-Rollesfon, 1996). In sub-Saharan Africa pastoralists like the modern Masai were above all reliant on cattle, and, forced southwards by increasing aridity, they spread progressively across that continent between 5000 and 1000 Cal. yr BP (Smith, 1980; Denbow and Wilmsen, 1986; Robertshaw, 1989). Like the plough, true pastoral nomadic societies such as these were restricted to the Old World prior to European expansion *c*.AD 1500, although a kind of agropastoralism based on llamas and alpacas did exist in the Andes during Prehispanic times (McGreevy, 1989). Pastoral nomads occupied land that was at best climatically marginal for settled farming and where the environment was inherently prone to great fluctuations in grazing, water, and hence food supplies. Livestock herds not only provided foodstuffs such as milk and meat, but also enabled pastoralists to move relatively swiftly to areas where grazing was seasonally available. Despite cultural practices such as warfare designed to evade dependence on the environment, livestock numbers were traditionally held in check by the environmental constraint of carrying capacity.

Pastoral nomads have had a distinctive relationship with settled agricultural societies throughout the later Holocene. This has often been of a reciprocal kind involving trade and transport, for example of salt in the Sahara. At other times, however, the relationship has been more hostile, with desert tribes attacking, overthrowing and replacing existing rulers of states and civilizations. In climatically marginal rain-fed farming areas of the Old World, the settled area expanded and contracted as centralized political stability alternated with weakness and the threat of nomadic incursions. In some cases the stimulus to out-migration by pastoralists from their homelands was a prolonged run of dry years in which pasture for their herds shrank and pressure on resources increased. Ellsworth Huntington (1915) saw much of history as linked to the beating 'pulse of Asia', with climatic cycles pushing the Huns, Mongols and other nomads outwards to the civilized rim of Eurasia. However, warfare is not the only possible response to animal carrying capacities being temporarily exceeded. At other times pastoralists have simply sold off livestock to reduce the imbalance, or settled down in adjacent land to become farmers.

Expansion at the Periphery

In addition to cultural developments such as pastoralism and urban civilization, the later part of the Holocene witnessed the last great primary colonizations of the globe, involving oceanic islands and the Arctic.

Conquest of the Northlands

The Arctic is one of the most hostile environments for human occupation and it required specialized cultural adaptation for it to be opened up successfully. As the ice sheets melted away during the early Holocene the modern tundra environment developed north of the boreal forest zone, bordering the seasonally frozen Arctic sea. H-f-g sites of Pleistocene age are known from Alaska and Northern Yukon; this was, after all, the gateway into the Americas from East Asia (see chapter 3), but the Arctic north was subsequently bypassed in mainstream cultural developments. In the second half of the Holocene, however, an Arctic maritime culture and technology developed that was specially adapted to the harsh and strongly seasonal periglacial environment (Dumond, 1980). Archaeological groups such as the Dorset and Thule cultures were the ancestors of the modern Inuit (Eskimo), and they survived with the same tool-kit of snow block houses (igloos), stone lamps, kayaks and harpoon hunting of seal and fish. The colonization was aided by the delayed Holocene climatic optimum in the Arctic, around 4500–3000 Cal. yr BP (Andrews et al., 1981). On the other side of the pole in northern Eurasia, cultural adaptations were based around reindeer herding, a way of life inherited from Pleistocene hunters, rather than specialized maritime fishing and hunting.

The first major European colonization of the North Atlantic region was carried out by Viking (Norse) settlers who reached Faroe, Iceland, Greenland and eventually Newfoundland in America between AD 850 and 1000 (1150–1000 Cal. yr BP). Unlike most oceanic islands, those in the North Atlantic possessed no endemic species at the time of the first long-ship landings. There had been a near-complete cover of ice over these land areas during the last glaciation, which left few, if any, refugia for plants and insects. The biota had, instead, to recolonize the islands at the start of the Holocene, either blown in on the wind or carried as – very cold! – passengers on ice floes drifting across the ocean (Buckland and Dugmore, 1991; Sadler and Skidmore, 1995). None the less, by the time of the Norse arrivals, Ári the Wise, writing in the twelfth century, was able to record that Iceland was forested with birch trees 'from mountain to sea shore'. Iceland soon became home to many new human settlers and their sheep, along with numerous stowaways such as fieldweeds and grain beetles, which had hitched a lift across the ocean. Within two to three hundred years there were over four thousand farms; a number comparable to that at the start of the twentieth century (Gerrard, 1991). Human impact on the Icelandic landscape since the initial *landnám* phase has been dramatic, with vegetation clearance and overgrazing leading to wind and water erosion of the unstable volcanic soils. Volcanic tephra layers provide convenient marker horizons with which it is possible to calculate how much sediment was transported and redeposited for different time intervals. Erosion seems to have started in upland areas, especially on grasslands and heathlands, and only later spread onto the low-lying woodlands (Gerrard, 1985; Dugmore and Buckland, 1991).

However, the human relationship with nature was no oneway train in this land of ice and fire. In AD 1783–4, for instance, a 30 km-long fissure eruption at Laki led to the loss of an estimated 24 per cent of Iceland's human and 70 per cent of its animal population, partly because the resulting volcanic gases led to excess fluorine intake (Thordarson and Self, 1993; Edwards et al., 1994; Grattan and Brayshay, 1995). The inhabitants of the North Atlantic islands also suffered severely from the climatic deterioration of the '**Little Ice Age**' (see chapter 7). The indigenous Inuit coped far more successfully with this climatic deterioration than the Vikings, who were obliged to abandon their Greenland settlements by AD 1500 (McGovern, 1980). These and other climatic changes can be indirectly reconstructed from the incidence of winter sea ice, which has been recorded in Icelandic chronicles for over a thousand years (Grove, 1988). The combined effects of population and grazing pressure, a variable climate and fragile soils explain why Iceland's landscape today is a denuded and barren – if spectacular – one.

The Pacific

The colonization of the Pacific was, if anything, an even more remarkable achievement than that of the Northlands. Prior to 3800 Cal. yr BP human settlement was restricted to the westernmost islands of Oceania, but in two great surges between then and 1000 Cal. yr BP virtually all of the Pacific came to be settled by seafaring peoples (Bellwood, 1980, 1987). The first surge occurred 3800–3200 Cal. yr BP associated with the expansion of the Lapita culture over Melanesia and western Polynesia (see figure 6.8). After a pause of almost a thousand

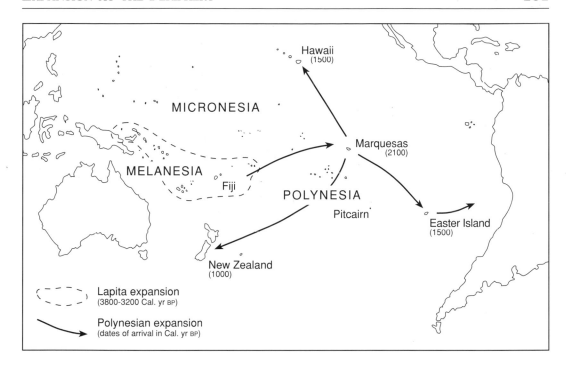

years, a second surge forward took place between 2100 and 1000 Cal. yr BP, involving the ancestors of the modern Polynesians. ^{14}C dates on early Polynesian sites indicate that colonization initially pushed eastwards from Fiji to the Marquesas before splitting three ways: north to Hawaii, east to Easter Island and – probably last of all – south to New Zealand. Genetic analysis of prehistoric human skeletons from Easter Island, Hawaii and the Chatham Islands (near New Zealand) has shown that all share a single of mitochondrial DNA lineage (Hagelberg et al., 1994). This confirms that these peoples shared a common origin, despite their enormous geographical spread, and that they passed through a population bottleneck at some stage in their colonization of remote Oceania. The Polynesian navigators may even have reached South America to bring back the sweet potato (*Ipomea batatas*) and – if Thor Heyerdahl is to be believed – the idea for the giant stone statues built on Easter Island.

There is no doubt that most of these were deliberate, purposeful colonizations and not the result of chance landings by sailors who lost their course. The long voyages must have been organized in advance for they took plants and animals as well as people to new islands. Sharing canoe space were dogs, chickens and pigs, and seeds or tubers of crops including coconut, taro, yam, banana and breadfruit. These formed the main cultivated foodstuffs of island horticulture, which were

Figure 6.8 The late Holocene colonization of the Pacific

locally supplemented by wild resources, particularly by fishing. The Pacific Islanders knew how to 'harvest' the sea's bounty, and it is probable that the sea acted more like a bridge than as a barrier between the island archipelagos.

Following founding of settlement, the standard portable economic package diversified to fit the resource base available on each island. On large, ecologically diverse islands such as Hawaii, agriculture became highly productive – even involving irrigation. This often came to support a hierarchical society whose tribal chief had as much power over land and life as did any English lord of the manor. By contrast, on smaller and more remote islands things had a tendency to get lost! The Lapita tradition of pottery making, for instance, died out on most Pacific islands, producing the unusual archaeological sequence of aceramic layers overlying ceramic ones. In many cases canoes and seafaring deteriorated so that inter-island contact declined and each archipelago became a virtual 'island universe'. Perhaps the ultimate example of cultural reversal occurred on Pitcairn, where Polynesian settlement died out completely (Diamond, 1994). Consequently, when the mutineers from the Bounty landed here in AD 1790 they found coconuts and breadfruit growing, stone platforms and statues, but no people. Small wonder that these became known as the Mystery Islands!

Just as islands provide 'living laboratories' for studying ecological processes, so they also provide controlled conditions for assessing human impact on natural ecosystems. The Pacific Islands were devoid of human population until the late Holocene colonization, although it should be noted that they lacked indigenous animal populations too, unlike most 'normal' ecosystems. The consequences of the arrival of Melanesian-Polynesian settlers on island ecosystems that had previously been untouched by human hand were far reaching. New Zealand, which was colonized by Polynesians at about the same time as the Norse reached Iceland, provides an informative example.

New Zealand has a moist sub-tropical to cool temperate climate which would seem ideal for forest growth. Yet when Captain James Cook and the other Europeans arrived here they found that almost half of the land was treeless. A proportion of this lay above the tree line, notably in South Island, but the majority comprised lowland scrub, fern or tussock grassland (see figure 6.9). However, within the grassland soils could be found abundant charcoal and even wood, suggesting that this had not always been open country. Indeed, pollen analyses have since shown that prior to the arrival of Polynesian settlement, only a limited area of central Otago with low rainfall failed to support continuous lowland forest cover. Polynesian

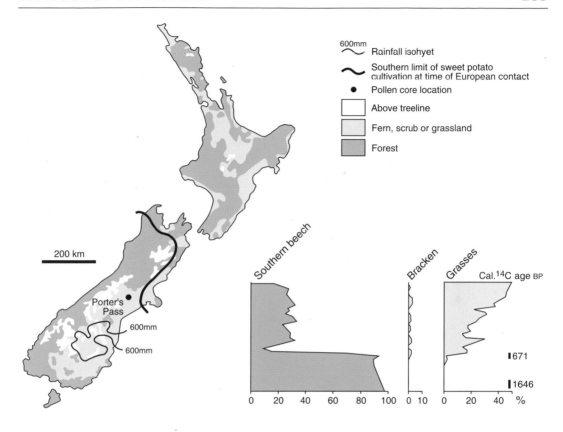

Figure 6.9 (Top) Extent of forest in New Zealand *c.*AD 1850; (bottom) pollen record of deforestation from Porter's Pass (after McGlone, 1983)

deforestation began in 1000 Cal. yr BP to become most severe between 800 and 600 Cal. yr BP, but stabilized some two centuries before European contact (McGlone, 1983; McGlone et al., 1994). Clearance was effected mostly by burning and does not seem to have been solely or even primarily because of the need for agricultural land. Maori horticulture was concentrated in North Island, where it was warm enough for the main staple crop – sweet potato – to be grown. Yet it was in South Island that deforestation was most severe. This was doubtless partly due to the low rainfall in eastern South Island, which made *Nothofagus* (southern beech) forest especially sensitive to fire and slow to recover after burning.

Clearance was almost certainly related primarily to other forms of resource use than farming, notably the desire to encourage the growth of bracken (*Pteridium aquilinum* var. *esculentum*), whose rhizomes were an important source of food. Burning may also have been associated with Polynesian hunting of the flightless moa birds (Dinornithiformes), whose densest populations lay in dry forest and scrub on the eastern side of South Island. The largest of these, *Dinornis giganteus* (plate 6.3), would have weighed 200 kg or more. 'In the absence of

Plate 6.3 Restored
moa bird along with
two Maori 'hunters'
(in fact medical
students dressed up
for the part), Dunedin,
1903

any other large wild animals,' as Richard Owen noted (1849,
p. 270), 'the whole art and practice of the chase must have
concentrated on these unhappy, cursorial birds.' And so it
proved. Faced with the loss of their habitat through burning
and ruthless human predation, moas and several other types
of bird were pushed into extinction only a few centuries after
the arrival of the Polynesians (Davidson, 1984; Cassels, 1984;
Anderson, 1989). Over-predation of seal and bird populations,
destruction of the forest and consequent soil erosion left a
degraded landscape and an impoverished economy. Well before European contact, South Island Maori society had gone
into eclipse.

An even more poignant example of human impact occurred
some 6500 km away on Easter Island. This, the most isolated
piece of inhabited land in the world, is renowned for numerous gigantic figures carved in stone and originally set in rows
on stone platforms or *ahu* (see plate 6.4). They were the work

of Polynesian settlers who first arrived there some 1500 years ago, probably benefiting from strong westerly winds which replace the normally dominant easterlies immediately before a major **ENSO** event (Caviedes and Waylen, 1993). Pollen cores from three crater lakes have shown that at the time of the settlers' arrival Easter Island was largely covered by palm forest (Flenley and King, 1984; Flenley et al., 1991). Although the forest had to be partially cleared to allow garden horticulture, timber continued to be a vital resource providing dug-out canoes on which fishing depended, wood for dwellings, palisades and fuel. It is almost certain that the stone statues could not have been moved and erected without the aid of supporting wooden timbers (Bellwood, 1987). The initial founding population grew so that between 850 and 300 Cal. yr BP (AD 1200–1650) the island possibly supported as many as 7000 people. This was the major period of monument construction and appears to have been associated with a society increasingly obsessed with rivalry and warfare. As demands on the finite timber resources of Easter Island increased, so trees were felled until eventually, as pollen records show, the palm forests were removed completely. On an island only 25 km long it would have been possible to see the last palm cut down

Plate 6.4 Easter Island statues. Pollen evidence records how the statue builders completely deforested the island

knowing that there would be no more. Following this ecological crisis, the first Europeans encountered a barren, almost treeless island littered with toppled giant statues from a by-gone era. The Easter Islanders had, it appears, destroyed the resources on which their society depended.

Of course, not all Polynesian impacts were as dramatic or injurious as on Easter Island. Many indigenous agricultural systems in the Pacific, such as integrated taro beds, became superbly adapted to local environments. Nevertheless, enough has been said to dispel the European myth of the South Sea Islands as an egalitarian paradise in which people and nature lived in perfect harmony. The Polynesian impact on island ecosystems of the Pacific had been a severe one.

Mediterranean Ecosystems

Mediterranean ecosystems represent a very different but equally striking type of environment. Part of their fascination lies in aesthetic appeal, but it also has much to do with the complex and often long-established pattern of human occupation and use of mediterranean lands. Also, any environment that is capable of producing most of the world's great wines is surely worthy of attention! Climatically, it is possible to distinguish five main areas of the world with summer drought, winter rain of cyclonic origin and a mean annual temperature of 15 ± 5°C. Central Chile, South Africa, South Australia, California and the Mediterranean basin itself all lie towards the western or southern margins of continental land masses at around 35° latitude. Their floras are physiognomically and ecologically similar, containing drought-tolerant forms such as the ever-green oaks of the Californian chapparal and the Mediterranean macchia. Mediterranean climates have also produced broadly comparable plant formations, ranging from steppic grassland at one extreme to sub-humid forest at the other (see table 6.3).

Of the major plant formations found in mediterranean-type environments only summer-dry evergreen forest, scrub and dry heath are distinctive to that ecotype. Sub-humid forest and steppic grassland exist elsewhere and cannot provide diagnostic evidence for the former existence of mediterranean-type ecosystems. In the circum-Mediterranean area, diagnostic taxa include evergreen oak, pistachio and olive (*Olea europea*), along with manna ash (*Fraxinus ornus*), hop hornbeam (*Ostrya*), phillyrea (*Phillyrea latifolia*), box tree (*Buxus*), strawberry tree (*Arbutus unedo*) and some species of pine. Some of these plants have **sclerophyllous**, drought-resistant leaves which are highly inflammable, so that mediterranean forests are prone to fires, whether natural, accidental or intentional.

Table 6.3   Mediterranean-type plant formations

Formation	Life-form	Height (m)	Understorey	Examples
Sub-humid forest	broad-leaved deciduous or coniferous trees	5–>30	perennial grasses + herbs; sclerophyll shrubs on poorer soils	montane forest (e.g. fir); deciduous oak-chestnut
Evergreen forest	broad-leaved evergreen trees (not all conifers)	5–30	perennial grasses + herbs; sclerophyll shrubs on poorer soils	forests of littoral pine, olive, carob, etc.
Open scrub	stunted evergreen trees + sclerophyll shrubs	2–8	perennial grasses, herbs + chenopods	macchia, chaparral, matorral
Open-scrub heathland	evergreen shrubs	0–2	–	garriga, malle heathland
Steppic grassland	perennial grasses + herbs	0–1	–	Mediterranean steppe; tussock grassland

Source: di Castri and Mooney (1973)

In addition to climate and vegetation, mediterranean-type environments are also distinguished by their soils (e.g. red *Terra Rossa*) and by their stark, often memorable landscapes. The image of the bare white limestone and blue sea of an Aegean island comes easily to mind. However, landscapes are not only the product of geology and climate but also of human agencies, direct or otherwise. To what extent is that bare white limestone a product of anthropogenically induced soil erosion? What has been the role of fire in shaping mediterranean plant communities? Indeed, how far are macchia and garriga natural vegetation formations at all, as they are associated so closely with human disturbance of the natural environment? According to Oliver Rackham's (1982) work in Greece, the characteristic Mediterranean scrub and heathlands are only maintained by cultural factors, such as grazing by sheep and goats. If ecological succession were allowed to proceed rather than being retarded and held as a **plagioclimax**, macchia would evolve to become full woodland while steppic grassland rather than heath would become the dominant non-tree vegetation. Pollen records from around the Mediterranean basin certainly show a marked increase in evergreen trees and shrubs during the later Holocene at the expense of sub-humid forest (see figure 6.10), although how much this is due to human impact and how much to climatic change is a subject much debated (Reille and Pons, 1992; Magri, 1995).

Comparison with the vegetation records in other parts of the world helps to clarify this issue. Of the five mediterranean-type regions, four have experienced major human modification only during the era of European colonization (i.e. post-AD 1500), but the fifth – the Mediterranean basin itself – has a complex record of cultural–environmental relations stretching back to early Holocene times. This provides an opportunity to compare across time and space those mediterranean ecosystems with short (less than 500 years) and long (more than 5000 years) records of agricultural impact. In fact, both California and southern Australia, although relatively free of human disturbance for the first 95 per cent of the Holocene, none the less supported sclerophyll evergreen communities such as chapparal scrub and mallee heath over thousands of years before European contact, proving that these vegetation communities can exist essentially free of human influence.

In the Mediterranean basin, evidence of early agricultural impact is hard to distinguish because Neolithic impact often occurred before vegetation had stabilized after the last glaciation. But once complex societies emerged around 4000 Cal. yr BP, vegetation disturbance becomes clearly visible in pollen diagrams (Bottema, 1982; Bottema and Woldring, 1990). Societies such as Bronze Age Greece were associated with the development of the Mediterranean 'triad' of wheat, olive and vine, along with the cultivation of other trees such as walnut (*Juglans*) (Zohary and Spiegel-Roy, 1975; Bottema, 1980). Tree crops came to be domesticated not by seed selection, but by other techniques of propagation, such as grafting. Unlike annual crops, cultivated trees have not extended much beyond their original climatic ranges, so that, for instance, olive groves remain a hallmark of Mediterranean-type landscapes. Arboriculture is a prominent feature of the so-called 'B.O.' (Beyşehir Occupation) clearance phase, which is present in pollen diagrams from southwest Turkey (Bottema and Woldring, 1990) (see figure 6.11). In several pollen diagrams this clearance phase starts just above a volcanic ash layer from the Thera eruption (Roberts et al., 1997b), and lasted from 3200 to c.1250 Cal. yr BP (1250 BC–AD 800). In this case the forest subsequently regenerated, but its species composition was changed. Prior to clearance the forest had been a mixture of pine, cedar, fir and deciduous oak, while afterwards it was dominated by pine alone. This may have been because forest grazing by livestock prevented regrowth of several tree species. Alternatively, it may be related to soil erosion during the clearance episode, with pine being able to survive better than other trees on the eroded, nutrient-poor soils (see figure 6.7B).

Palaeoecological records show that the intensity and timing of human impact on 'natural' forest vegetation varied from one part of the circum-Mediterranean region to another. The magnificent cedar forests of the Middle Atlas mountains of Morocco, for example, have been shown by pollen analysis to have continued unbroken right through the later Holocene (Lamb and van der Kaars, 1995). Human impact has come relatively late here, and until the nineteenth century was focused in Morocco's lowland zone (Flower et al., 1989; Lamb et al., 1991; McNeil, 1992). In other areas, forests were modified and used as a managed resource for grazing, woodfuel and other products, rather than deforested; the *dehesa* system for using the southern Spanish oak woodlands, for example, has been shown by archaeology and pollen to go back to Bronze Age times (Stevenson and Harrison, 1992). Elsewhere in the Mediterranean, the late Holocene saw a major decline in fir forests, in some cases being replaced by beech and hornbeam (Reille et al., 1996; Watts et al., 1996; Magri, 1997). In the uplands of Corsica beech forests were managed for pig pannage, but it is not always clear why anthropogenic action should have led to the changes that occurred in the species mix of Mediterranean forests, if indeed it did. It follows that deforestation was by no means the only way in which people transformed the native Mediterranean woodlands, and that changes through time in the ratio of tree to non-tree pollen only tell us part of that story.

Mirroring the modifications made to natural vegetation, soil erosion has become widespread around the Mediterranean basin during recent millennia. Evidence for this comes both from eroded hillsides and from valley alluviation. Many Mediterranean river valleys contain an alluvial deposit of historical age, which Claudio Vita-Finzi (1969b) has labelled the 'Younger Fill'. It contains pottery and other dateable remains which indicate that the main period of deposition was from late Roman to early medieval times (*c.*1700–500 Cal. yr BP). Mineral magnetic studies have also been able to trace its source to eroded catchment topsoils, rather than to stream incision of bedrock (James and Chester, 1995). But while the soil eroded from hillslopes helped to form new and fertile coastal plains, as at Troy (see figure 6.5), it also choked the mouths of many rivers (Kraft et al., 1977; Bintliff, 1981). The classical harbour at Ephesus, for example, had to be abandoned because of siltation, and now lies 4 km from the sea (see plate 6.5) (Eisma, 1978).

Andrew Goudie's (1993) comment that 'in many cases of environmental change it is not possible to state without risk of

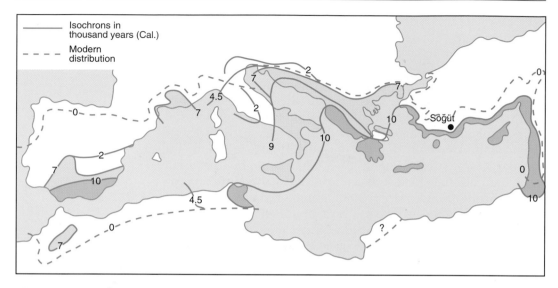

Figure 6.10 Holocene spread of summer-dry woodland in the Mediterranean basin (data from Huntley and Birks, 1983, and other sources)

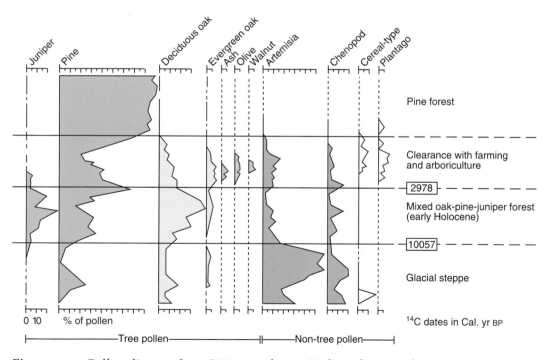

Figure 6.11 Pollen diagram from Söğüt, southwest Turkey, showing clearance phase (based on van Zeist et al., 1975)

Plate 6.5 Classical harbour of Ephesus, now 4 km from the sea due to historical river alluviation

contradiction that it is man rather than nature which is responsible', certainly applies to the origin of the Younger Fill. In particular, there has been debate about whether soil erosion and deposition were brought on by climatic change or by human land abuse (Wagstaff, 1981; Chester and James, 1991; Ballais, 1995; Provensal, 1995). The fact that alluviation began before 3000 Cal. yr BP in some areas (Davidson, 1980; van Andel et al., 1986, 1990) suggests that climatic oscillations were unlikely to have been solely responsible. On the other hand, it would be wrong to assume that all eroded Mediterranean hillsides are anthropogenic landscapes. A salutary lesson in this comes from the badlands of southeast Spain. Here integrated geo-archaeological studies have found *in situ* Bronze Age sites lying uneroded on vulnerable bare slopes (Wise et al., 1982). The badlands had already formed by *c.*4500 years ago, and later agricultural clearances have done little to alter erosion rates. While giving the appearance of recent degradation, these and similar Mediterranean badlands, such as Cappadocia in Turkey (see plate 6.6 on p. 194), are ancient landscapes created by long-term geological instability rather than the ravages of the goat and the plough.

In most cases historical soil erosion was a combined product of natural and cultural forces. Mediterranean lands typically have thin or erodible soils, steep slopes, a vegetation vulnerable to fire, and rainfall that can be intense and erosive.

These natural factors made mediterranean ecosystems fragile under human impact. The replacement of forest by terraced vineyards and olive groves acted to conserve the soil, and so long as this agricultural system was maintained land degradation was avoided. But this was only a metastable equilibrium state. When the Roman Empire was invaded by eastern nomads, rural security declined and soil terraces fell into disrepair. The metastable equilibrium was upset and in many cases severe soil erosion followed, with valleys becoming choked by sediment. This helps to explain why the Younger Fill was usually formed after and not during the classical era of maximum land utilization.

Historical and palaeoecological research in the Mediterranean, as in many other areas, is increasingly based around integrated projects using a range of techniques – from pollen analysis through to archaeological survey – to reconstruct regional landscape history (e.g. Barker, 1995a, 1995b; Jameson et al., 1994). These multi-disciplinary projects reveal most Mediterranean landscapes to have been the result of human impact – direct and indirect – upon a fragile environment. The aromatic herbs, cork oak (*Quercus suber*) and bare white limestone that are considered to be the embodiment of the natural Mediterranean world have become important features only in late Holocene times. As the plagioclimax formations of macchia and garriga have increased, so there has been a depletion of sub-humid forest, notably of deciduous oak, cedar and fir (Pons and Quezel, 1985). On eroded hillsides it is doubtful whether these plant formations could ever return.

The Making of the Landscape: the British Isles

The landscapes of Britain and Ireland have long been a source of fascination, partly because of their diversity, but also because their development is so deeply rooted in history. Those landscapes have been produced by the interaction of the peoples of the British Isles and their natural environments, now and in the past. But how is the history of that interaction to be reconstructed?

One approach has involved using the landscape itself as a source of information. Celtic, Roman, Saxon and Norman – these and other periods created their own cultural landscapes, which were superimposed on pre-existing ones to create a palimpsest of elements of different age and origin (Hoskins, 1955). Distinctive features such as the feudal open field system of the English Midlands enable the landscape to be decoded and 'read like a book'. An account of the ecological history of the British Isles based upon elements still extant in the landscape might proceed as follows.

> The natural, primaeval vegetation cover of lowland Britain was dense oak woodland except on sandy or calcareous soils, where there was beechwood, grassy downland or lowland heath. By contrast, much of highland Britain and Ireland was covered by moorland or blanket bog. These highland wildscapes have undergone much less cultural modification than lowland landscapes, at least up until twentieth-century coniferization. The main clearance of native woodland occurred in the Roman or in the early medieval periods. Prehistoric peoples, though leaving their imprint on the landscape in the form of stone circles and barrows, were few in number and their impact small in comparison with historical times.

In fact, most if not all of an account such as this would be incorrect. This is because landscape-based reconstructions are able to cope much better with recent than with distant time periods. Very little of those earlier prehistoric and primaeval landscapes escaped the shipwreck of time, and the few surviving relicts are not always representative of what was formerly widespread. Because information shrinks rapidly backwards through time, historical myopia is created – vision is clear for things chronologically close, but blurred for those further away.

The same is also true of a second source of information – documentary evidence. How, for instance, can Norman and Celtic contributions to landscape change be realistically compared, when one is well documented historically and still evident in the landscape, while the other is largely buried in Prehistory and whose traces have largely been reduced to hilltop ditches and embankments? Fortunately, one documentary source does provide an extremely important base-line datum, for the lowland zone of the British Isles at least, from which landscape studies can work backwards. That source is the Domesday survey of AD 1086. This survey shows clearly that nine centuries ago lowland England was already densely settled, with the Weald and the Forest of Dean as two of the few remaining areas of wildwood. Over most counties of eastern England and also in Devon and Cornwall, woodland accounted for less than 5 per cent of the total land area. Oliver Rackham (1980, ch. 9) has estimated that only 15 per cent of Domesday England as a whole was wooded, less than half the area then under arable land. Although the Normans were responsible for reducing the woodland cover still further, from 15 per cent to 10 per cent by AD 1350, it is clear that the vast majority of primary forest clearance had taken place before the Norman

Plate 6.6 Ancient
eroded badland,
Cappadocia, Turkey

conquest in the eleventh century. If it is to be established
when and how the 'natural' landscape of around 6000 years
ago came to be the largely cultural one recorded in the Domesday Book, the mainly non-documentary evidence of archaeology and palaeoecology has to be considered (Jones, 1986).

The primaeval forest

The notion of an original pristine British landscape is something of a myth, for even in the early Holocene, Mesolithic
hunter-fisher-foragers were capable of significant ecological
impacts (see chapter 4). The nearest equivalent to a primaeval,
naturally vegetated landscape would be the forests of the
Atlantic period, between 8000 and 6000 years ago. Pollen
diagrams show that in the Atlantic period the only substantial
areas without forest were the 5 per cent or so of the British
Isles above the tree line, which lay at around 500 m (see figure
6.14). There were then trees on what are today bare and wind-
swept landscapes like the remote Shetland Islands to the north
of Scotland (Bennett et al., 1992). Even coastal wetlands periodically supported trees, as is shown by submerged beds of
pine, oak and yew (*Taxus*) uncovered from Fenland peat
(Godwin, 1978).

The composition of the forest varied significantly across the country, forming a polyclimax vegetation (Bennett, 1989). Over most of the highland zone, woodland was dominated by the sessile oak (*Quercus petraea*). In Scotland around the Great Glen lay the Caledonian pine forest, a relict of the pinewoods that had briefly covered most of the British Isles early in the Holocene. Elsewhere in the highland zone, local variants of the flora included birchwoods in the far north of Scotland and elmwoods in the middle third of Ireland. Blanket bog had started to develop in some parts of Ireland and Wales but still only formed a minor component of the landscape. The character of lowland zone woodland was equally varied, although those variations were generally more subtle in the way they merged with one another. Despite the oft-quoted 'mixed oak forest' label, oak was by no means the dominant tree species. In many parts of lowland England, such as East Anglia, limewoods (*Tilia*) were extensive (Greig, 1982), while alder (*Alnus*) occupied the damp ground in river floodplains and around lakes. Hazel (*Corylus*) was a widespread understorey shrub throughout the British Isles, and locally it formed substantial trees. Beech, although present, did not become an important component of woodland until the Bronze Age.

By the time Neolithic agriculture reached the British Isles around 5700 Cal. yr BP, eustatic sea levels were within 2 m of the present day and the coastline had its familiar modern configuration. Consequently, the first farmers had also to be familiar with sailing in order to cross the English Channel and the Irish Sea with their new domestic crops and livestock. As elsewhere in northwest Europe (see chapter 5), the initial impact of agriculture was relatively slight, clearances of the forest being localized and temporary. Forest fallow agriculture, where it operated, created a mosaic of secondary woodland in various stages of regeneration. The forest cover was therefore largely maintained but its species composition altered; elm, for instance, declining in importance after around 5700 Cal. yr BP. Early Neolithic agriculture – like Mesolithic hunting and foraging – involved modification of existing woodland ecosystems rather than their complete replacement by agro-ecosystems. If anything, the impact of agriculture decreased in the middle and later Neolithic (5200–4500 Cal. yr BP), when it is thought there was a widespread and extended phase of woodland regeneration (Whittle, 1978).

Neolithic and early Bronze Age society reached its apogee between 5000 and 3800 Cal. yr BP with the completion of great monuments at Silbury Hill and Stonehenge. These are not only impressive feats of engineering but also testimony to the existence of periods of labour surplus within the prehistoric economy. The main legacy in the landscape of Britain's first farmers is consequently not their farms or villages (a part of the economic process), but their burial and ceremonial places (a part of the social process). The obvious importance of social activities should make us cautious in explaining Neolithic or early Bronze Age impact on flora or fauna solely in terms of agricultural economics. For example, land snails and pollen from buried soils at Avebury, Silbury Hill and Stonehenge show that the chalk landscape had by then already been changed from woodland to open pasture or scrub (see figure 6.12; Technical Box VII). However, the area around these sites may have been kept clear by grazing animals so that the monuments remained visible for astronomical observations or similar non-economic reasons. Evidence of permanent pasture around ceremonial and burial sites need not mean that, for example, shifting agriculture was not taking place elsewhere in the landscape.

The period 3800–2600 Cal. yr BP saw the end of this older society and the beginning of Celtic Britain. From both social and environmental points of view, this shift was fundamental. Celtic society was – or at least it became – a hierarchical, tribal one and its mode of production became progressively less dependent on domestic subsistence agriculture. The asso-

Technical Box VII: Mollusca analysis

TECHNICAL BOX

Molluscs are small invertebrate animals with an external shell made of calcium carbonate. Mollusc shells are well preserved in alkaline soils and sediments, from which they can be extracted without chemical pre-treatment. Identification is generally to species level. The species composition of land snail assemblages is largely dependent on climate and on local habitat, especially vegetation cover. During the terminal Pleistocene and early Holocene land snails found in colluvial and similar sequences closely mirror vegetational changes over the same time period (Kerney, 1977; Preece, 1993). Later in the Holocene land snails have proved especially valuable in archaeological contexts, such as soils buried beneath burial mounds (Evans, 1972, 1993; Thomas, 1985). Species which are shade tolerant or intolerant provide good indicators of whether the vegetation was formerly wooded or open, and hence of forest clearance and regeneration (see figure 6.12). The stable isotope analysis of mollusc shells has also proved to be a means of reconstructing past changes in rainfall in arid environments (Goodfriend, 1992).

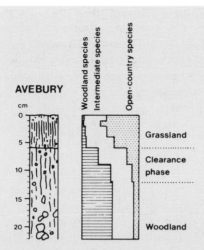

Figure 6.12 Land snail record of woodland clearance, Avebury, southern England (after Evans, 1972)

Molluscs are also found in aquatic and marine environments, as bivalves like mussels as well as gastropods (snails) (see plate 6.7). Most freshwater molluscs, having wide climatic tolerances, instead provide an indication of local aquatic conditions; for example, whether the water was moving or stagnant (Ložek, 1986). Marine shells are above all found at the seashore. Their incorporation in coastal sediment sequences has made them an invaluable tool in studies of past sea-level change (Richards, 1985).

Plate 6.7 Freshwater mollusc shells *Dreissena* (bivalve) and *Lymnaea* (gastropod)

ciated demographic changes were equally profound. Peter
Fowler (1983, p. 33) has estimated that Britain's population
rose from about 14 000 to 2 million people between 4000 and
2000 years ago, the latter figure being higher than that recorded in the Domesday survey. These changes were manifest
in the landscape in the creation of organized arable field systems and other forms of land allotment, and defensive 'hill
fort' settlements (Bowen and Fowler, 1978). Although ard-marks
are known from Neolithic contexts, it was after 3600 Cal. yr BP
that the use of the ard (or scratch plough) appears to have
become widespread. Animals were no longer raised solely for
meat, but also came to provide traction for ploughing, manure
to maintain arable soil fertility, transport, wool and milk
(Sherratt, 1979). New crop strains were also adopted, notably
spelt wheat (*Triticum spelta*) and the celtic bean (*Vicia faba*).
In other words, after an era of social investment in ceremonial
monuments, spare productive capacity was switched to the
more pragmatic task of creating Britain's first large-scale agricultural landscapes (Fowler, 1983, p. 36).

The period from 3800 to 2600 Cal. yr BP also saw an important extension of settlement in many parts of highland
Britain. A good example is provided by Dartmoor, where large
parts of the Bronze Age landscape have been preserved (Price,
1993).

Shaugh Moor: a Bronze Age landscape

Southwest Dartmoor is today an area of china clay workings,
and from the late 1970s a rescue project took place to investigate a 5 km² block of landscape before it was destroyed by the
bulldozer. The Shaugh Moor project from the outset involved
not only archaeological excavation and survey but also studies
of soil and vegetation histories in order to establish the natural
environment of the Bronze Age. The integrated approach employed at Shaugh Moor allowed local spatial variations in environmental conditions to be tied directly to patterns of past
human land use.

Shaugh Moor incorporates a series of low stone walls – or
reaves – some of which are linked to the wider system of
prehistoric land boundaries that covers the whole of the Dartmoor uplands (Fleming, 1978, 1983), while others form smaller
areas of enclosed pasture (see figure 6.13A). Although animal
bones were not preserved under the acid soil conditions, hoof-
prints of sheep and cattle were found in a Bronze Age reave
ditch, and pollen of pastoral weeds was much more abundant
than that of arable species. Carbonized cereal grains were
virtually absent and phosphate levels do not indicate manuring within the enclosed fields. The economy was therefore

A

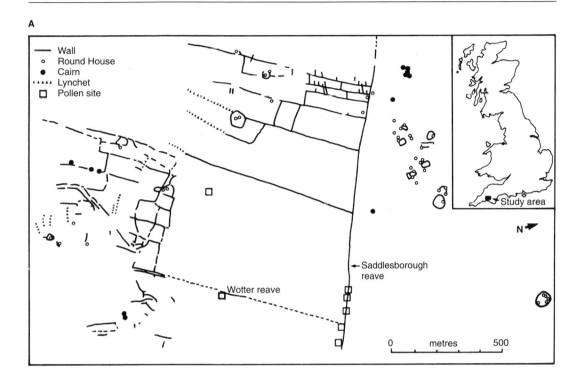

B

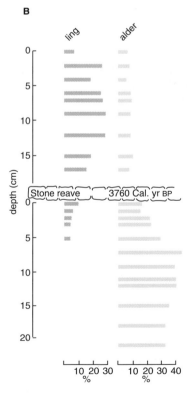

Figure 6.13 Shaugh Moor, Devon, England: **A** relict Bronze Age cultural landscape (based on Smith et al., 1981); **B** pollen diagram from above and below Wotter reave (based on Balaam et al., 1982)

evidently based on pastoralism rather than arable farming. Settlement evidence included round, stone-built farm dwellings, which may only have been occupied seasonally, perhaps as part of a transhumant pattern of land occupation. Both dwellings and reaves involved extensive use of wooden posts, indicating the availability of timber and its exploitation. [14]C dates on wood charcoal and peat show that the reaves were built around 3800 Cal. yr BP and that they and the settlements were used for about 1200 years.

Pollen diagrams were studied from peat deposits and soils below and above Bronze Age reaves. These record Neolithic clearance and subsequent regeneration of woodland prior to reave construction. As archaeological remains of Neolithic age are hardly recorded at Shaugh Moor – presumably because of their subsequent destruction in the Bronze Age – this pollen evidence serves as a reminder of how selective cultural landscape preservation can be. By 3800 Cal. yr BP vegetation was a mosaic of alder woods, hazel scrub on the better-drained slopes, and acid grassland on the plateau top above about 300 m. After a millennium of Bronze Age pastoral occupation, most of this had been transformed into moorland, with ling (*Calluna*), other heathers and acid grasses growing on peaty stagnopodzols (see figure 6.13B). Soil changes from acid brown earths to podzols and gley soils had already commenced when the Bronze Age occupation began, but under it, degradation continued so that by 2600 Cal. yr BP both soils and vegetation were indistinguishable from those of today. Shaugh Moor was virtually abandoned at this time, leaving behind a Bronze Age landscape disturbed only by subsequent mining for tin and china clay.

The environmental impact of permanent agricultural clearance

The climatic deterioration around the Sub-Boreal/Sub-Atlantic transition was critical to the human ecology of Britain's highland zone. The Bronze Age had marked the 'high tide' of upland farming in the British Isles, but in the face of an increasingly hostile climate and deteriorating soil and vegetation resources, the upland population retreated to lower ground. The environmental changes that prompted this abandonment were as permanent and damaging to human welfare as the desertification occurring in Africa today. And like African desertification, these changes were due at least partly to human misuse of a fragile environment. Dartmoor's prehistoric pastoralists had almost literally been treading on dangerous ground.

But if the Bronze Age 'enclosure movement' created a landscape of environmental dereliction in parts of northern and western Britain, it had precisely the opposite effect over most of lowland England. Here, landscape reorganization was associated with an increase in agricultural productivity and carrying capacities. The forest fallow system of Neolithic cultivation had allowed most of the woodland cover to be maintained, but with 'enclosure' came permanent clearance of the forests of the lowland zone. Quantitative estimates of the regional extent of forest clearance are difficult to obtain from pollen evidence, but the later Bronze and pre-Roman Iron Age (i.e. 3000–2000 Cal. yr BP) was probably the most active period of wildwood destruction in English history. Woodland clearance of those parts of highland Britain and Ireland not under peat and moorland came slightly later, in most pollen diagrams between 2500 and 1500 Cal. yr BP (see figure 6.14).

From Julius Caesar's description that 'the population [of southern Britain] is exceedingly large, the ground thickly studded with homesteads' (*De Bello Gallico*, v, 2), it certainly appears that the Romans took over a landscape in lowland Britain that was already largely agricultural. This is confirmed from aerial photographs, which show dense Iron Age settlement, for instance, on the gravel terraces and floodplains of major rivers such as the Thames (Jones, 1986). At the margins of their empire, the Romans may have been the first people to undertake major woodland clearance. The construction of Hadrian's Wall – to keep out the Scots – used large numbers of mature oak trees (McCarthy, 1995), and pollen diagrams from peat bogs along the wall show a dramatic change from a mainly wooded to an open landscape at about this time (Dumayne and Barber, 1994). Despite their different social bases, Celtic, classical and early feudal modes of production were all overwhelmingly agrarian in character (Anderson, 1974; Arnold, 1984). The post-Roman period witnessed a decline in population and woodland regeneration in some areas, but the basic pattern of land occupation established in the pre-Roman Iron Age was not greatly altered. It seems that the distribution of wooded and non-wooded land and the territorial divisions of most estates and parishes recorded in Domesday England had already then been in existence for at least a thousand years, and possibly twice that long (Taylor, 1983).

The subsequent survival or destruction of woodland in medieval times lay almost entirely in the hands of the dominant social class, notably as Royal Forests and deer parks. In contrast to prehistoric and early historic times, there was no longer common access to woodland resources. Woodland had formerly been less valuable than other forms of land use, but after AD 1250 the value of trees and fuelwood rose steadily,

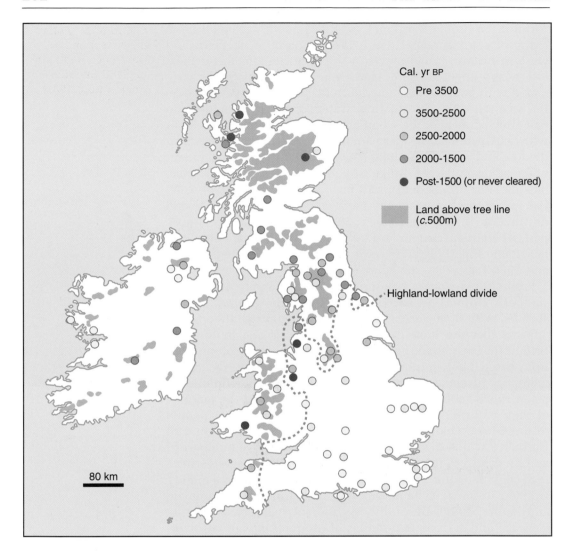

Cal. yr BP
○ Pre 3500
○ 3500-2500
◑ 2500-2000
◔ 2000-1500
● Post-1500 (or never cleared)

Land above tree line (c.500m)

Highland-lowland divide

80 km

Figure 6.14 Map showing dates of first major forest clearance phase in Britain and Ireland from pollen and land snail evidence (data from Turner, 1981; Berglund et al., 1996 and other sources). Note the uniformly early date of initial clearance across lowland England, and the more varied ages in the highland zone

providing landowners with a good income as well as the pleasures of the hunt (Rackham, 1980, p. 170ff.). Woodland conservation and management made economic sense, and forests were ruthlessly protected against 'wood stealers'. Damage to hollies or thorns – which offered safety to young oak trees – attracted up to three months' hard labour with a whipping in each month, for example. The conservation ethic did not apply equally to rich and poor, on the other hand. In 1251, King Henry III's Christmas dinner included 730 deer, 1300 hares, 395 swans, 115 cranes and 200 wild boar – the boar coming from the last remaining population in England which became extinct soon afterwards (Rackham, 1980, p. 181).

Pollen records tell an essentially similar history of land use to those from archaeology and documented history. This can

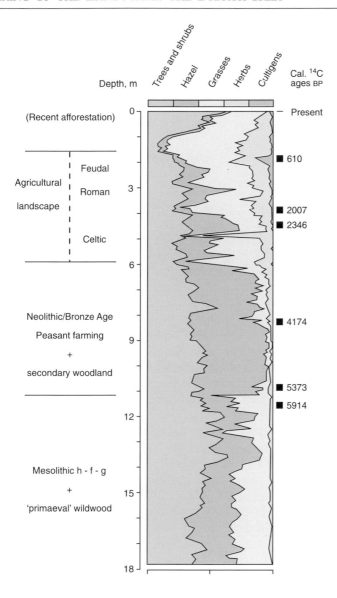

Figure 6.15 Pollen diagram from Rimsmoor, southern England (modified from Waton, 1982)

be illustrated by a diagram from Rimsmoor in Dorset which, in many ways, serves to summarize the whole history of human impact on Britain's lowland zone vegetation during the Holocene. Rimsmoor is a small bog lying near to the edge of the chalk, and contains no less than 18 m of pollen-bearing sediment. As a result it contains a detailed record of mid-to-late Holocene ecological change, with a pollen count every three to five years for key sections of the diagram (Waton, 1982). Prior to 5700 Cal. yr BP the vegetation comprised a stable mixed deciduous woodland community, with oak, elm, ash, lime and alder as the main tree taxa (see figure 6.15). Abundant charcoal indicates periodic Mesolithic burning of the landscape. Between 5700 and 3200 Cal. yr BP hazel was abundant, oak fluctuating, herbs and grasses rare, but with cereal pollen occasionally present. This is what might be expected under a system of extensive, forest fallow shifting agriculture, with much of the land under regenerating secondary woodland or scrub. Modification of woodland species composition is indicated by a decline in elm (c.5700 Cal. yr BP) and lime (c.4200 Cal. yr BP).

Clearance associated with early 'Celtic' landscape organization around 3000 years ago is very clearly marked. There is a sharp increase in herbs and grasses including those, such as *Rumex* and ribwort plantain (*Plantago lanceolata*), typical of agricultural activity. Over the next two millennia the intensity of land use appears to have been broadly constant, despite periodic changes in the balance between pastoral and arable components of the landscape and the introduction of some new crops, such as rye (*Secale*) and hemp/hop (*Cannabis*) (see also Edwards and Whittington, 1992). Cultivation reached a peak of intensity in about the fourteenth century, and subsequently declined, with reafforestation by pine in the eighteenth and nineteenth centuries. In short, this pollen diagram records three phases of human land-use activity prior to the advent of modern times, each one lasting about 2500 years. The first was hunting and gathering under wildwood; the second, small-scale peasant farming within secondary woodland; and the third phase dominated by an agricultural landscape of fields and farms created under complex, stratified societies.

The later Holocene woodland clearance recorded at Rimsmoor and most other pollen sites had permanent consequences for catchment soils. The original brown earths have in some cases been preserved beneath Neolithic and Bronze Age barrows and earthworks, and they are quite different from the soils that now surround them on the chalklands. In particular, the fertile but superficial cover of loess has been almost completely eroded to leave the present thin calcareous soils (Catt, 1978). The eroded soil has 'sludged' downhill to form extensive colluvial deposits at the base of slopes (Bell, 1983). Some of this was moved into river systems, whose sediment load was increased further by clearance of alder woodland on lowland floodplains (Brown, 1988). These influxes of sediment led to the widespread accretion of fine-grained floodplain alluvium after c.3000 Cal. yr BP (Buckland and Sadler, 1985; Brown and Keough, 1994), while upland rivers underwent more complex cycles of river incision and aggradation (Macklin et al., 1992). On the chalkland, erosion of loess was not normally sufficient to prevent tree regeneration. But in other environments, pedogenic changes had far-reaching consequences for the total ecosystem that prevented it reverting to its original state. One such extreme example of irreversible ecological change is provided by the limestone plateau of the Burren in western Ireland. The Burren's thin soil cover, which was none the less able to support pine, yew and birch woods during the Holocene, has been almost totally eroded down karstic fissures during the recent millennia (Drew, 1982; Watts, 1984). All that is left is bare limestone pavement incongruously criss-crossed by Bronze Age fields with no soil inside them (see plate 6.8)!

Conclusion

It is easy to fall into the trap of describing human impact on the natural world solely in terms of 'degradation' and 'impoverishment', especially when considering issues such as soil erosion or deforestation. In fact, agriculture – at least in its pre-capitalist form – has generally been an agent of ecological diversification. It caused the relative homogeneity of primaeval forest ecosystems to be replaced by a mosaic of habitats, from downland to heath, macchia to garriga. These are all semi-natural plagioclimax ecosystems, created and maintained by human action, and their fates came to be intimately associated with particular modes of agricultural production. The English chalk downlands, for example, were maintained as a grassland ecosystem only because of a long-standing system of sheep grazing. Once this traditional form of land use no longer operated, the natural processes of plant succession began to allow annual grasses to be replaced by scrub and eventually to revert to woodland. The 'ecological clock' was in this case held up by another domestic herbivore – the rabbit – but this proved a temporary check when myxomatosis decimated the rabbit population in the 1950s. Consequently, conservation of the springy turf and panoramic views of England's downs will not be achieved simply by protecting that environment from human disturbance.

The long interplay between human production and different natural environments often makes it effectively impossible to separate natural from cultural influences. All landscapes created by agricultural societies are intrinsically both natural and cultural. Of course, this makes it difficult to pinpoint simple cause-and-effect mechanisms for environmental changes such as soil erosion or blanket bog formation. But it would be artificial and illusory to separate what was the result of inter-action between people and nature. The environmental changes that occurred during the period from the mid-Holocene until 500 years ago cannot be neatly and succinctly summarized; they were too diverse through time and across space. Diversity

Plate 6.8 Former field systems on the Burren, western Ireland, now containing limestone pavement and devoid of soil. Behind this Cashel (round enclosure) at Cahercommaun is hazel woodland, a relict of that which was formerly extensive over the Burren

was to give way to uniformity towards modern times, however, although due not to a natural mechanism such as climate but to a cultural one. The creation of the European world system and industrial capitalism, and their impact on the natural world, form the focus of the next chapter in this story.

CHAPTER SEVEN

The Impact of Modern Times (500–0 Cal. yr BP)

Introduction

The last half-millennium of the Earth's natural history has been a time of dramatic and accelerating change. One has to look to the beginning of the Holocene, with the climatic amelioration after the last ice age and the Neolithic agricultural revolution, to find a period which produced changes of comparable significance for human–environmental relations. What makes the last five centuries so distinctive is, of course, the quantum leap in human impact on nature that has occurred during this time. Climate and other natural agencies have been far from static over the last 500 years, but there are no grounds for believing that they operated in any way differently from earlier in the Holocene except where prompted by human agency. The justification for treating the period since AD 1500 as a separate time bloc within the Holocene is therefore solely based on culturally induced changes in the natural environment.

The 'Industrial Revolution' is usually considered to be the most dramatic economic and cultural shift to have taken place in modern times. With the advent of industrial capitalism, human use of resources became increasingly exploitative and most daily productive activities became separated from land, climate and the rest of nature (Quaini, 1982, p. 122ff.). The countryside and wilderness became the escape from routine labour, not part of it. Even agriculture has become industrialized through mechanization, and land-use activities are now determined more by decisions taken in Brussels or Washington than locally in Galloway or Michigan. For many of the world's peoples and ecosystems, however, the dominant change of recent centuries has been brought about not by industrial capitalism but by European colonial expansion (Arnold, 1996). European culture encountered new and unfamiliar natural environments overseas, and what resulted was unlike either had been before the encounter. Adaptations were often harmonious, as in the vineyards of California, but more than occasionally were not; the dustbowl of the western American prairies providing a classic example of maladaption of people to what was – for them – new land.

Industrial capitalism and European colonialism are both products of the last two centuries, but the shift to a modern pattern of human–environment relations began before this. A third event had commenced three centuries earlier, when Vasco da Gama rounded the Cape of Good Hope and Columbus set foot on the New World. In AD 1450, Eurasia, the Americas, sub-Saharan Africa and Australasia were unknown to each other, but by AD 1550, European explorers had brought all except the last within a system of global contact (Wallerstein, 1974). The explorers returned not only with gold but also with a plant to be 'smoked' and fine bird feathers for ladies' hats. The discovery and subsequent interchange of plants and animal domesticates brought to an end the discrete crop cultures that had existed for so many thousands of years in western and eastern Eurasia, and in the Americas (see chapter 5). Indeed, so radical was this transformation that such things as Irish potatoes, Indian palomino ponies, Australian sheep and Canadian wheat are now accepted as native. In fact, all were introduced and adopted following the age of European discovery.

For plant crops there was two-way traffic between the Old and New Worlds, but the situation was different for animals. The Americas had almost completely lacked indigenous domestic animals before AD 1500, and even the wild mammal fauna was species-poor following the wave of megafaunal extinctions at the end of the last glaciation. The ecological potential for intrusive animal species was immense, and the colonization of the New World was consequently accomplished by four-legged as much as by two-legged Europeans. The omnivorous domestic pig was a particularly effective 'primary colonizer' and was followed by cattle, horses and sheep, some of which escaped to become **feral** species. In the southwest United States, for example, donkeys (*Equus asinus*) used during the nineteenth-century gold rush returned to the wild and are now more numerous than the native pronghorn antelope (*Antilocapra americana*). The feral donkeys are able to outcompete the antelope by virtue of their consumption of a much wider range of graze and browse, and are seriously degrading the vegetation as a result.

It has often been the unintended rather than the intended consequences of European contact that have had the most far-reaching ecological effects (Crosby, 1986). Species that had evolved separately over millions of years suddenly came into contact to produce a homogenized, melting-pot biota (Buckland et al., 1995). Weedy plants such as plantain (*Plantago*), familiar in European pollen diagrams from the Neolithic on, expanded their range to disturbed habitats throughout the temperate world. Some native species benefited too; ragweed (*Ambrosia*) was one such in North America, and the *Ambrosia* pollen rise is a standard indicator of clearance associated with the advance of settlement across that continent. But just as some creatures profited, so others lost out. Before European contact millions of passenger pigeons (*Ectopistes migratorius*) were to be found in the woodlands of eastern North America, and millions of buffalo (*Bison bison*) grazed upon its prairie grasses and herbs (McDonald, 1981). By the end of the nineteenth century, the buffalo population of the entire continent had been reduced to a few hundred head (see plate 7.1), and of the passenger pigeon there was neither sight nor sound,

Plate 7.1 The North American bison, numbered in their millions before European expansion, were almost shot to extinction with the coming of the railroad

a bird shot out of the sky as its habitat was reduced through forest clearance.

Extinction rates were highest on island ecosystems, especially on those few which remained uninhabited prior to European contact (Steadman et al., 1991). As Charles Darwin noted on his *Beagle* voyage, isolation produces unusual, even bizarre, island adaptations, including giant flightless birds like the dodo. Once that isolation is broken and competition introduced in the form of pigs or rats carried on board ship, native island biotas are ill adapted for survival, and many species became as dead as the dodo. It has been argued that it was with observations of resource depletion and species extinction on island ecosystems such as St Helena that modern western conservation philosophy has its beginnings (Grove, 1995).

The consequences of abrupt external impact are well illustrated by the sub-tropical Atlantic Islands of the Madeiras. Madeira possessed no indigenous human population when it was discovered and colonized by the Portuguese in the fifteenth century AD, but it did have an almost complete cover of 'great trees', including the sharp cedar (*Juniperus oxycedrus*) and the extraordinary and long-lived dragon tree (*Dracaena draco*). Indeed, the very name that Madeira was given meant 'timber'. To the settlers, however, the trees were an impediment and 'it was therefore first of all necessary . . . to set fire to them. By this means they razed a great part of the forest' (Cadamosto, 1455; quoted in Naval Intelligence Division, 1945, p. 46).

Following the conflagration – supposedly seven years long – cattle, sheep and rabbits introduced by the Portuguese prevented woodland regeneration so that today the remnants of the original forests are confined to a few precipitous ravines on the north coast of the main island. On the smaller island of Porto Santo the environmental impact following contact was even more dramatic. Furthermore, it can be traced to a single event: the release onto the island of a female rabbit and her offspring, taken to the island by the father-in-law of Columbus, Bartholemeu Perestrello. In the absence of local predators, the rabbits bred prolifically and 'overspread the land, so that our men could sow nothing that was not destroyed by them' (Crosby, 1986, p. 75). Despite attempts by the settlers to kill them off, the rabbits devoured so much of the vegetation and crop cover that severe erosion followed, water sources dried up and the now degraded island had to be abandoned for human settlement. The impact of human arrivals extended to many elements of the island ecosystem. For example, the native land snail fauna is – even today – rich and highly endemic, but none the less it has been significantly depleted through extinction during the last five centuries (Goodfriend et al., 1994).

One final feature of the time period covered in this chapter that needs mention is the sources of information available; specifically, the wide availability of sources other than those provided by proxy methods. No one would think to reconstruct modern European history entirely from archaeological finds, although it is worth remembering that in some parts of the world written records go back no further than the early twentieth century AD. It was formerly assumed that the same also applied to environmental history. That is, palaeoecological data were thought to be superfluous for recent centuries because documented historical ecologies were available instead. In consequence, palaeoecologists tended to ignore the uppermost layers of their peat or lake sediment cores and instead concentrated their efforts on events further back in time. Work instigated since 1970, notably by Frank Oldfield (1977, 1983), has shown that even for historical times, proxy sources can be crucially important. In particular, many contemporary environmental problems such as 'acid rain' benefit from an historical perspective but lack adequate documented records. Palaeoecological methods can be applied to environmental changes over decades as well as over millennia, although it has presented some new challenges, notably in dating techniques. Even so, it would be foolish to ignore archival and similar sources of recent environmental history (Hooke and

Kain, 1982; Sheail, 1980). Indeed, it is often when used in combination that historical, archaeological and palaeo-environmental records provide the fullest insights into how and why people have transformed the natural world around them (Roberts and Butlin, 1995). This chapter does not rely solely on proxy data, therefore, but uses them in tandem with other types of information.

Climatic Changes in Historical Times

It is common experience that no two years have the same weather. Further, years of 'exceptional' weather can run to-gether to form more prolonged periods of 'abnormal' conditions, such as have caused the droughts that have affected Africa's Sahelian zone since the 1960s. But our expectation is that these deviations are random events and that over decades and centuries they will cancel each other out to give an essentially stable climate. Testing the idea of underlying climatic stability is not without its difficulties. Many proxy climatic indicators such as pollen-based vegetation changes or cycles of stream erosion are amenable to alternative interpretation as we move forward into historical times. There is also the thorny question of whether the global climate has been warming as a result of a culturally induced rise in atmospheric CO_2 levels (Barnett et al., 1996).

To help disentangle these different factors, a longer-term perspective may be of value. Has climatic stability been the norm over periods of less that 500 years during the Holocene? The answer must be that it has not. Secular climatic cycles generally lasting between 200 and 600 years are recorded as recurrence surfaces within peat bog profiles (see figure 6.2, p. 163), via salinity fluctuations in non-outlet lakes (Fritz et al., 1991; Laird et al., 1996), in cores from ice caps (Thompson, 1991; Meese et al., 1994) and by many other palaeoclimatic indicators. Some of these archives, such as variations in tree-ring widths (Bircher, 1986; Briffa et al., 1990), can be resolved to individual years. This permits investigation of the potential causes of short-term changes to the climate, such as ENSO events, injections of dust into the atmosphere from explosive volcanic eruptions, and variations in solar activity linked to sunspot cycles. Holocene environmental history consequently warns us not to expect climate to have been constant over the last five centuries.

Specific evidence for recent climatic change comes from environments near to ecologic or geologic thresholds, such as the upper tree line in mountains. Indeed, montane regions like the Alps provide rich pickings for climatic reconstruction. In

addition to biotic indicators such as tree-rings, alpine glaciers are climatically sensitive and leave behind a stratigraphic record of their former extent (see chapter 2). Moraines mark recent glacier advances downvalley (see plate 7.2), and dating by dendrochronologies and early landscape paintings shows that their retreat to modern positions began only in the mid-nineteenth century (Rothlisberger and Schneebeli, 1979). This forms the end of a prolonged period of relatively adverse climate in northern and central Europe – the so-called Little Ice Age (Lamb, 1977; Grove, 1988). Conventionally, the Little Ice Age began *c*.AD 1590 and ended *c*.AD 1850, during which major rivers like the Thames froze over on many occasions (plate 7.3). On the other hand, it would be wrong to imagine that the whole of the intervening period was cold and damp. Nor should it be thought that the climatic deterioration was anything like so severe as during the last Ice Age proper; the drop in mean temperature was only around 1°C, not the 10°C of Pleistocene glaciations. Even so, the shift in the mean climatic conditions over those two and a half centuries was sufficient to have measurable ecological and human consequences, especially in upland environments.

New tree growth was prevented in marginal areas and while mature trees were often able to survive, they were stunted and marked by narrow annual growth rings. Conditions of upland peat bog development were similarly affected. In warmer, drier

Plate 7.2 Little Ice Age trim-lines and moraines (marked), Val d'Hèrens, Swiss Alps. The two glaciers were formerly joined, and as late as AD 1859 they extended downvalley to beyond the point from where the photograph was taken

THAMES FROZEN OVER
11ᵗʰ FEBRUARY
1895

Plate 7.3 The River Thames frozen over in 1895. Earlier during the Little Ice Age, frost fairs were held on the solid ice that covered the river surface

climatic phases such as the early medieval warm episode of the eleventh to thirteenth centuries AD, peat bog growth had been predominantly in the form of dry hummocks. With the shift to a cooler, wetter climate on the other hand, hummocks gave way to 'wet lawn' and eventually to *Sphagnum*-dominated, permanently wet pools. Stratigraphic analysis of peat profiles in northwest England by Keith Barber (1981) has shown a close correspondence between climatic oscillation and different phases of peat development over the last thousand years (see figure 7.1B).

Adverse climatic conditions were inimical to domesticated as well as wild plants, and crop failure in Europe's upland valleys meant that times of feast were replaced by times of famine (Ladurie, 1972). Hill farmers in northern and western Britain no longer found themselves within the agro-climatic limit for cereal cultivation, and much marginal land was abandoned for settlement between AD 1600 and 1750 (Parry, 1978). Similarly, it is hard not to see a causal link between the medieval cooling of the fourteenth and fifteenth centuries and the extinction of the Norse settlement in Greenland around AD 1500. Climatic deterioration brought not only poor harvests

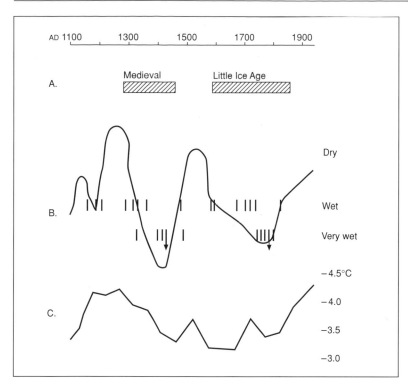

Figure 7.1 European climatic records over the last nine centuries: **A** alpine glacial advances; **B** surface wetness curve, Bolton Fell moss peat bog (after Barber, 1981); **C** reconstructed winter temperatures for southern England (after Lamb, 1981)

and pasture, but also a higher risk of geomorphological hazards including landslides, avalanches, glacier outbursts and other floods (Grove and Battagel, 1981). The human consequences of floods, frosts and droughts were often recorded in official reports, diaries and other documents, and these now form valuable sources of information about historical variations in climate in parts of the world such as Iceland (Ogilvie, 1992). It can, in fact, become temptingly easy to use climatic fluctuations as a catch-all explanation for socio-economic changes in early modern times, when in reality non-environmental factors were at least as important. Societies were perfectly capable of 'bucking' the climatic trend under appropriate circumstances. The period of the Little Ice Age in southern Turkey, for example, actually saw a major shift of population from plains into montane valleys, rather than the other way round. The political and economic instability of life in later Ottoman times more than offset any climatic adversity in the decision to move into the safety of the Taurus mountains.

Most evidence for the human consequences of secular climatic variations over the last 500 years comes from temperate areas of the world, such as Europe and China, with rich historical records available to draw on (Bradley and Jones, 1992). But to judge by twentieth-century experience, drought-prone regions can be expected to have been at least as severely affected by any climatic fluctuations, especially those in rainfall. Greater aridity may have marked the North American Great Plains at the time of the Little Ice Age, to judge from lake-salinity variations (Fritz et al., 1994). Increased dust flux onto the Quelccaya ice cap in the Peruvian Andes suggests a similar tendency towards drought during this time interval (Thompson et al., 1986). Elsewhere in the Americas, drought phases seem to have coincided with warming phases, such as the Medieval Warm Epoch before c.AD 1300, rather than cooling ones (Stine, 1994). Similarly in Africa, Sharon Nicholson (1980) used a range of evidence to conclude that at the time of the Little Ice Age, rainfall on the southern (Sahelian) margin of the Sahara was significantly above that immediately before or since then. Stratigraphic records of lake levels show that Lake Chad was 4 m above the mid-twentieth-century mean from AD 1570 to 1750, while in the two to three centuries prior to this both Lake Chad and Lake Bosumtwi in Ghana were at low levels (Talbot and Delibrias, 1977). This drier phase appears to have been important in causing a retreat of settlement on the inland delta of the River Niger (McIntosh, 1983). Similar hydrological and climatic fluctuations in recent centuries are recorded in the water-level history of Lake Abhé in Ethiopia (Fontes et al., 1985).

Even more dramatic evidence comes from Africa south of the equator. Sediment cores from Lake Malawi have shown that this, one of the largest and deepest of all the world's lakes, ceased overflowing and its water level fell by almost 100 m (Owen et al., 1990). Although this great regression occurred before the start of documented historical data in the mid-nineteenth century, it is recorded in oral traditions and is dated from cores to between c.AD 1450 and 1850. Calibrated climatically, this fall represents a rainfall level only 50–70 per cent of twentieth-century values, and would have created a drought whose severity and duration make recent climatic events in Africa pale in comparison. It is clear, then, that recent centuries have been every bit as turbulent climatically in the tropics as they have been in Europe, with its better-known 'Little Ice Age'.

If variations in climatic and other natural factors have been far from insignificant in recent centuries, what of changes to the environment brought about by human agency, specifically by the export of European farming to the New World, Australasia and Africa?

Land-use History and Soil Erosion

Land degradation through accelerated soil erosion has occurred on a number of occasions during the last 10 000 years, for

example as a result of abrupt climatic change or fire disturbance. However, moving towards the present day it is human-
induced acceleration of soil loss that becomes the dominant
influence. In particular, the progressive adoption and intensification of agricultural modes of production have led to widespread conversion of natural woodland or grassland vegetation
into farming land. Of course clearance for crop cultivation
does not leave the soil completely bare. Some of the original
flora may remain, for example in hedgerows, while the crop
itself affords considerable protection from raindrop impact.
The invasion of weeds during fallow periods or of secondary
forest after abandonment by tropical shifting agriculture has a
similar protective effect. On grazed grassland the ground cover
may hardly be reduced at all. None the less, agricultural clearance has certainly increased soil erosion losses well above the
long-term geological norm (Douglas, 1967).

The environmental consequences of soil loss from agro-
ecosystems are complex and highly variable (Thornes, 1987;
Boardman et al., 1990). Nutrient depletion and other forms of
soil degradation can crucially affect conditions for plant growth,
not only changing crop yields but also restricting the range
of wild plants that could eventually re-occupy the site. One
useful indicator of environmental impact is 'soil life', or the
balance between the rate of soil loss and the creation of new
soil by substrate weathering. Soils on alluvial floodplains, for
example, may take only 50 years to form and are generally
subject to low rates of erosion, making their soil life effectively infinite. In contrast, soils on the ancient land surfaces
of Africa have taken millions of years to form and in some
cases have lives as short as the rate at which they are being lost,
which can be as little as ten years on soils actively eroding
today (Elwell and Stocking, 1984). Nor is impact restricted to
the site of degradation. Eroded soils are washed downslope
and downstream to push up the sediment loads carried by
rivers and affect their aquatic life.

Human beings are not passive to land degradation, and conservation measures to maintain soil fertility are surely as old
as agriculture itself. Traditionally, tropical shifting cultivators
have left tree stumps in their fields and intercropped to reduce
the impact of the violent convectional rains on the soil surface
(see plate 7.4). Manuring, which provides nutrients and helps
maintain soil organic matter levels, is another very ancient
practice (Wilkinson, 1982, 1989). The laborious work of ter-
racing China's loess plateau by peasant farmers has greatly
reduced erosion in an environment where natural soil losses
were amongst the highest in the world. However, there may be
a long-term price to pay for cultivating marginal land such as
this. Soil conservation systems need maintenance, and if they

break down it can have environmental consequences worse than if the land had been left alone in the first place.

Over the past 500 years, European expansion and colonialism have brought many important changes in land use and traditional culture–environment relations. In some continents, such as North America, this involved wholesale transfer of European populations, with the consequent eclipse or demise of indigenous peoples. In other parts of the world, especially in the tropics, European contact left the indigenous peoples intact but disrupted their established mode of production and threatened their traditional relations with nature. 'You ask me to plough the ground. Shall I take a knife and tear my mother's bosom?' was the reaction of the Amerindian Smohalla to the introduction of Old World agricultural technology.

In order to establish the impact that European cultural contact had on the natural environment, case histories of the relationship between land-use history and soil erosion in different parts of the world are discussed here. The first example derives from 'Old Europe' and serves as a yardstick for comparison with nineteenth-century European settlement and clearance in the eastern United States, and twentieth-century 'colonial' impact in the highlands of Papua New Guinea. Lake-based studies provide the main source of proxy soil erosion data in these examples. Lake sediments are of particular value because the same cores used to reconstruct sediment influx and catchment erosion rates can also provide evidence of land-use changes via their pollen record (Oldfield and Clark, 1990). In considering these examples it is more important to contrast different trends through time than to compare the specific values computed for past erosion rates, for these are partly determined by local factors such as geology and lake-catchment size (see Technical Box VIII on pp. 221–2).

HAVGÅRDSSJÖN in southern Sweden is a relatively shallow, kettle-hole lake situated in an area of predominantly arable farming. Over recent centuries the main change in land use has been agricultural intensification and in particular the conversion of pasture into arable land. John Dearing calculated past erosion rates by taking a series of 1 m-long sediment cores set in a grid over the lake surface (see figure 7.3, Technical Box VIII). Dry sediment influx rates and mean catchment erosion losses have been produced for different time periods during the last 350 years (see figure 7.2). The main change recorded is an increase in the inferred erosion rate during the nineteenth century. Study of historical records has shown that this coincided with an expansion in the area of ploughed land and a sharp increase in the sheep population. Although this agricultural intensification appears to have increased soil losses from the catchment, it is significant that erosion rates were already

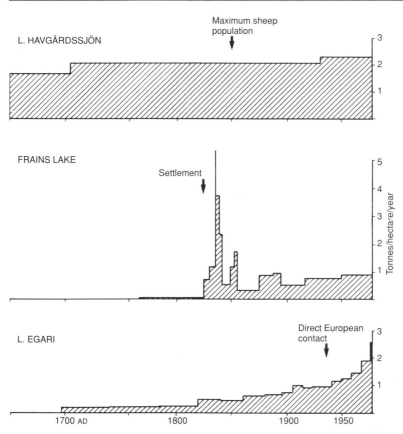

Figure 7.2
Comparative soil erosion rates reconstructed from lake sediments: Sweden, Michigan and Papua New Guinea (data from Dearing et al., 1987; Davis, 1976b; Oldfield et al., 1985)

relatively high even in the seventeenth century. This is surely related to the fact that clearance of the original woodland cover took place before the modern period, especially in Viking times (Dearing et al., 1987). A longer sediment core from the lake does show a progressive increase in the sediment accumulation rate from 5000 yr BP onwards, as do many other northwest European records (Dearing, 1994). At Braeroddach Loch in Scotland, sediment influx increased in a series of steps through time, notably at around 5000 yr BP with the arrival of Neolithic agriculture and around 350 yr BP with clearance of secondary woodland and land drainage (Edwards and Rowntree, 1980). In this catchment soil losses under recent agricultural land use represent a thirtyfold increase compared with that under early Holocene forest cover. Without doubt, land degradation has occurred in northwest Europe in recent centuries related to population growth and agrarian pressure (Blaikie and Brookfield, 1986, p. 126ff.; Dodgshon, 1988); the same pressure that contributed to the subsequent European expansion overseas. But land clearance and degradation have a long antiquity in this corner of the world, and records such as those from Havgårdssjön can only be understood by considering

Plate 7.4 Shifting cultivators in the tropics often leave the larger trees in place; their leaves intercept the erosive rainfall and their roots bind the soil together, so helping to reduce soil losses

changes in prehistoric and early historic times, as well as in the modern period.

FRAINS LAKE, Michigan, is somewhat smaller than Havgårdssjön, but has a similar lake : catchment area (z) ratio (see table 7.1). As in southern Sweden, the natural vegetation around Frains Lake would have been mixed temperate woodland, but in Michigan it was cleared for agriculture much more recently, mainly in the nineteenth century AD. How would the sudden arrival of European settlement affect erosion in the Frains Lake catchment? Margaret Bryan Davis (1976b) recon-

Technical Box VIII: Reconstructing past erosion rates

Lakes and valleys act as receptacles for materials which are removed from their catchments, the majority of which are deposited as alluvium, colluvium or lacustrine sediments (Dearing, 1994). In order to overcome the problem of differences in the rate of sedimentation over the surface of these 'sinks' it is necessary to take multiple cores or cut sections. These need to be correlated with one another by methods such as magnetic susceptibility (see figure 7.3). Using magnetic susceptibility profiles it is possible to establish which places have fast and which have slow sedimentation rates, and to select for detailed dating those with representative sediment profiles. In the case of a lake, accumulation rates can be established for different time periods in the past once cores have been dated and correlated. The mean rate of sediment accumulation over the whole lake bed (usually recorded as gm/cm^2/yr) may be used as a measure of the rate of sediment influx into the lake, which in turn can be converted into mean catchment erosion rate (t/ha/yr) (Dearing, 1986; Foster et al., 1988).

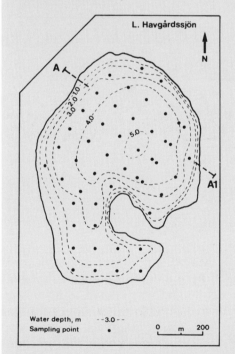

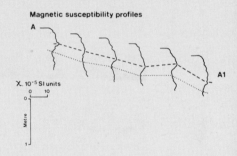

Figure 7.3 Reconstructing past erosion rates in a lake catchment (modified from Dearing, 1983)

This approach to environmental reconstruction uses sediment yield as a surrogate for soil loss, but because it cannot take proper account of material temporarily stored upstream, it is difficult to use it as a comparative measure of absolute erosion rates between catchments. A second problem is that with the exception of non-outlet lakes, even good sediment traps will incur some downstream loss of material, especially of solutes and fine particles such as clay. Third, sedimentary records produce an aggregated record of erosion and give no immediate indication of where within the catchment erosion took place. Fortunately, it may be possible to retrodict different sediment sources from the character of the deposit. In Loe Pool, Cornwall, for example, a layer of red clay recorded in sediment pools was found to be rich in haematite (Fe_2O_3) and tin (Sn). Its metalliferous character could be related to tin mining in the lake catchment during the 1930s (Coard et al., 1983). Although extrapolated catchment erosion rates increased to a mean of $c.4$ t/ha/yr at this time, it would seem to be entirely due to the point source where mine waste was discharged. It is unlikely that agricultural soil losses elsewhere in the catchment were any higher than their recent level of 0.2 t/ha/yr. Different magnetic properties can also be used to discriminate between, for example, material eroded from topsoil and from unweathered bedrock (Thompson and Oldfield, 1986).

Table 7.1
Comparative statistics for three lakes used to reconstruct erosion rates

	Havgårdssjön	Frains Lake	Lake Egari
Altitude (m above sea level)	51	260	1800
Lake area (ha)	55	6.7	8.3
Catchment area (ha)[a]	141	18.4	9.2
z (lake-catchment) ratio	0.39	0.36	0.9
max. water depth (m)	$c.5.5$	$c.10$	11
core density (per ha)	1	3.3	0.6
core correlation	m.s.	pollen	m.s., volcanic ashes

[a] excluding lake   m.s. = magnetic susceptibility

structed the land-use and erosion history from sediment cores which were cross-correlated by pollen analysis and sediment stratigraphy. The modern settlement was known to have been founded in AD 1830, and the associated forest clearance showed up as a clear *landnám* phase in pollen diagrams, with a sharp rise in ragweed (*Ambrosia*) pollen. The same horizon was also marked by a change in sedimentation from organic gyttja mud to inorganic clay, reflecting the inwash of minerogenic material following deforestation.

One result of her study was to show that in a small area at the centre of the basin, sediment had accumulated ten times faster than elsewhere on the lake bed. The fact that no less than 3.7 m of sediment could build up in a localized part of the lake in the 140 years after 1830 graphically illustrates the need to use multiple cores in any lake-based reconstruction of total sediment influx rates. Sediment core samples were burnt to produce ash weight from which the influx rate of inorganic material over the last 180 years was calculated (see figure 7.2). This showed a dramatic jump in erosion from around 0.1 to over 5 t/ha/yr with forest clearance, before stabilizing at a new equilibrium level at around 0.9 t/ha/yr once farming was established. Longer cores showed that sediment accumulation rates had been stable and low over the remainder of the Holocene, although a slight rise occurred after 2000 yr BP, possibly related to indigenous Indian land-use activities. In contrast to Havgårdssjön, the sequence from Frains Lake shows a tenfold increase in mean sediment yield in recent centuries, and rather than rising gradually, erosion rates rose abruptly upon European impact. Over the mid-latitude portion of eastern North America as a whole, lake sedimentation rates have increased threefold as a result of European vegetation clearance and other impacts (Webb and Webb, 1988). Perhaps the most significant result of the Frains Lake study was in identifying the importance of the unstable transitional phase between different land uses – in this case forest and farmland. In **ergodically** based studies (e.g. Wolman, 1967), retrodicted modern erosion rates under different land-use types failed to recognize the importance of transitional instability associated with the deforestation phase.

Similar sequences to that at Frains Lake have been recorded elsewhere in eastern North America. In a varved sediment record from Crawford Lake in southern Ontario, the impact of nineteenth-century European settlement on the mixed temperate forest can immediately be seen from pollen studies carried out by 'Jock' McAndrews (figure 7.4). In the uppermost zone (IV) there is a dramatic increase in the pollen of ragweed, grasses and plantain as the forests were cut down or burnt, the latter being reflected in a marked increase of charcoal in the lake sediments. Interestingly, this is not the first period of disturbance to the forest that is recorded at Crawford Lake, although it is certainly the largest. An earlier clearance phase between AD 1361 and 1650 (pollen zone II) was linked to native Amerindian settlement and maize cultivation around the lake (Finlayson and Byrne, 1975). The associated change in forest composition from beech to oak to pine dominance may have been related to the increase in forest fires resulting from burning by Iroquois Indians (Clark and Royall, 1995; Clark, 1995). Alternatively, it may have been a product of Little Ice Age cooling (Campbell and McAndrews, 1993).

Grace Brush (1994), working at Chesapeake Bay, has also linked environmental changes recorded in sediment cores to a

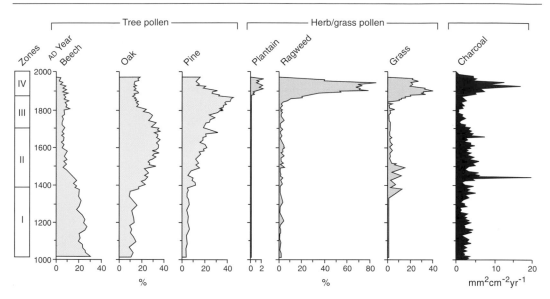

Figure 7.4 Pollen and charcoal record for the last 1000 years from Crawford Lake, Ontario, Canada (after Campbell and McAndrews, 1993; Clark and Royall, 1995)

Plate 7.5 Sweet potato gardens at the forest margin, highlands of Papua New Guinea (2700 m)

Time span (Cal. yr BP)	Catchment erosion (tonnes/ha/yr)	Land use
5700–3750	0.12	Initial forest clearance
3750–1000	0.09	Swamp margin irrigation
1000–500	0.20	Further clearance with dry land horticulture
500–200	0.61	Intensified horticulture with sweet potato
200–present	0.60	Further use and European impact

Table 7.2 Estimated erosion rates at Yeni Swamp, Papua New Guinea

Source: Gillieson et al. (1986)

cooler, damper Little Ice Age climate, in this case involving a slow-down in sedimentation in the centuries prior to c.AD 1786. Alternatively, the reduction of catchment forest fires and consequent lack of erosional disturbance along the eastern seaboard of the United States may have been linked to the massive reduction in Amerindian population caused by small-pox and other 'European' diseases (Roberts, 1989). Diseases emptied many American landscapes of their native populations well in advance of the main arrival of immigrants from across the Atlantic, and allowed forests that had previously been managed to regenerate (Whitney, 1994). The main wave of forest clearance by European settlers around Chesapeake Bay, as elsewhere, was marked by a sharp increase in the rate of sediment flux. Furthermore, Frank Oldfield has been able to use magnetic 'signatures' to separate the two main sources of sediment into this estuary, namely bedrock eroded along its cliffs and eroded catchment topsoils (Thompson and Oldfield, 1986, ch. 10). These components could be identified in sediment cores from the central channels of the estuary, which revealed that whereas bedrock formed the main sediment input before AD 1600, this was subsequently replaced by eroded topsoil, which became the main source of sediment by AD 1880.

The third study area is the HIGHLANDS OF PAPUA NEW GUINEA, and it offers both a different cultural context and natural environment. It experiences intense tropical rains and has steeply sloping, erodible soils, which have been protected from rainsplash erosion by dense montane rainforest. These highlands have been thought of as culturally isolated, being bypassed by nineteenth-century colonialism and only falling within a 'western' orbit during the 1930s. Over the last 50 years the traditional mode of production had been disrupted by the introduction of a cash economy. Severe soil stripping is evident in many parts of the highlands today, but it is unclear whether this is connected to twentieth-century contact or to a much longer indigenous process of agricultural intensification.

Agriculture began in the Papua New Guinea highlands as early as 9000 years ago (see chapter 5), involving both swamp and dry-land cultivation of yams, taro, sugar cane and bananas, and husbandry of pigs. Over the course of the Holocene it gradually intensified to support one of the highest population densities of any tropical swidden cultivators. There was, as Jack Golson (1977) put it, 'no room left at the top' by the time of European contact. What impact did intensive indigenous agriculture have on land degradation in this fragile environment?

Studies based on the sediment stratigraphy of swamp and cave deposits show a progressive increase in soil erosion rates during the later Holocene (see table 7.2). Soil losses were lowest under raised-bed wetland horticulture, and rose after 500 yr BP with the introduction of the sweet potato (*Ipomea batatas*). This crop was swiftly adopted on account of its wide environmental tolerance and quick growing time, and permitted a demographic growth and intensification of garden horticulture (see plate 7.5). Significantly, the sweet potato is a South American domesticate, probably brought to the Philippines by the Spaniards in the sixteenth century AD. The fact that it reached the highlanders of Papua New Guinea soon afterwards showed that their pre-European cultural isolation was by no means complete (Allen and Crittenden, 1987). Intensive garden cultivation may have led to soil nutrient depletion, which would provide an origin for the grasslands found in some parts of the highlands.

Although soil degradation did take place in the past, especially between 500 and 100 yr BP, erosion losses before European contact are less than those recorded after clearance in the much more robust environments of northwest Europe and eastern United States. The link between contemporary erosion and cash cropping would appear to be confirmed by lake-based studies of past erosion rates in the Papua New Guinea highlands (Oldfield et al., 1985). At LAKE EGARI sediment accumulation rates have been used to reconstruct erosion losses, showing a tenfold increase over the last 150 years, and doubling since the 1930s (see figure 7.2). Although there was no new settlement of land by Europeans as there was at Frains Lake, the environmental consequences of European contact have been similar in both cases. It is a remarkable fact that traditional swidden and wetland agriculture operated in the ecologically fragile highlands of Papua New Guinea for over 8000 years, eventually supporting almost a million people, without serious environmental degradation. This situation only changed when indigenous environmental relations were disrupted, first, with the introduction of a new exotic domesticate – the sweet potato – and second, with the advent of the twentieth-century cash economy.

Pollution Histories

If the hallmark of agricultural impact has been clearance of natural vegetation to make way for field or pasture, the hallmark of industry has surely been pollution. Industrial society has modified the cycling of natural elements to an extent that was never possible under previous agricultural modes of production. The use of fossil fuels has reduced the dependence of industrial society on biological energy fixation through green plants, and has profoundly altered the global carbon cycle as a result. Other elemental cycles that have been altered include those of key micronutrients such as phosphorus and toxins such as lead.

The term pollution implies disturbance of the natural environment by putting the wrong substances in the wrong quantities in the wrong places. Although pollution includes artificial elements and compounds, such as radioactive caesium-137 or toxic DDT, it more usually involves naturally occurring ones like sulphur or nitrogen. This makes it difficult to establish whether the present state of an ecosystem has been modified by pollution or not. Lake Erie is much richer in nutrients than Lake Superior, but is this because of sewage and industrial effluent from around Lake Erie's shores, or because the geology of the two lake catchments is different anyway? A second problem that bedevils pollution studies is that pollutants move downstream or downwind from their point of origin to affect other localities. Establishing the source of pollution, and consequently the cause of it, can be a tricky business. Palaeo-ecological techniques cannot overcome these obstacles on their own, but they do have some crucial advantages over other approaches (Smol, 1992). They provide longer and more continuous histories of water chemistry and biology than monitored records and smooth out the random variations of individual years. They further enable the stable base-line condition of the ecosystem to be established, if one ever existed.

In the remainder of this chapter the palaeoecological histories of some industrial-age pollutants and their consequences for biota are described. In particular, it is shown how palaeo-ecological techniques have been used to trace the origin of **eutrophication** and **acidification** in freshwater ecosystems.

Eutrophication: natural or cultural?

Ecosystems can be nutrient-rich (eutrophic) or nutrient-poor (oligotrophic), and can change naturally from one to the other. In lakes, eutrophication can occur through the natural ageing process as the lake is infilled with sediment and the volume of the hypolimnion is reduced (see Technical Box IV, p. 95). Eutrophication can also be culturally induced through the application and subsequent runoff of chemical fertilizers to farmland or by the discharge of sewage or industrial waste into streams and rivers. In all of these cases, cultural eutrophication is an unintended, indirect consequence of land-use, sanitation or industrial activities. Nutrients, particularly nitrate and phosphate, are washed away by runoff and carried downstream into rivers and lakes. Most organisms are subject to Liebig's Law of the minimum, by which productivity will be restricted by the availability of the substance that, in relation to the needs of each organism, is least abundant in the environment (Hutchinson, 1973). In aquatic ecosystems, nitrogen and phosphorus are usually the limiting substances that hold in check greater primary productivity in the form of algal growth.

At first sight it might appear that eutrophication should be a beneficial process, because it encourages greater biological productivity. Farmers, after all, apply nitrogen and phosphorus to their fields precisely in order to stimulate more prolific plant growth. Unfortunately, the downstream consequences are not always so desirable. First, many organisms are adapted to oligotrophic conditions; the natural habitat of trout, for example, is in clear if relatively infertile upland streams and lakes rather than in murky, eutrophic lowland waters. Second, blue-green algae, which are rather inedible by zooplankton, increase more rapidly with eutrophication than other algal forms. Third, eutrophication can go too far, to produce a hyper-eutrophic environment. Primary productivity may increase to the point where dead algae use up most of the oxygen as they decompose whilst sinking to the lake bed. Lake waters become de-oxygenated and higher organisms such as fish, instead of thriving on an abundant food supply, find themselves restricted to the upper few metres of the lake where they are not starved of oxygen. In some cases, such as many of the Norfolk Broads in eastern England, hyper-eutrophic conditions have eliminated stabilizing aquatic macrophytes such as *Chara* and *Najas* and their waters have become smelly and opaque (Moss, 1980, p. 284ff.).

The sediment profile of a presently eutrophic lake can be used to establish whether eutrophy has been the long-term, stable state of the ecosystem or whether it has eutrophied over time. In studies of recent lake history (less than 500 years), only the upper metre of sediment need normally be cored, making fieldwork a relatively easy task (see plate 7.6). One simple measure of former lake productivity is the organic matter content of the sediment profile. Under eutrophic conditions more organic matter is produced within the lake than is oxidized as it descends to the mud–water interface. On the other hand, an upcore increase in organic matter could be caused by changes other than eutrophication – by erosion of

Plate 7.6 Freeze core taken in annually laminated sediments from Baldeggersee in Switzerland's Alpine foreland. The change from light to dark sediments dates to AD 1885 and marks the onset of anoxia in the bottom waters of the lake brought on by cultural eutrophication (after Lotter et al., 1997)

peat from the lake catchment, for example. Organic matter changes alone cannot be conclusive. A second and perhaps more obvious measure is the chemical composition of the profile, notably its phosphorus and nitrogen content. This has been successfully analysed in cores from a number of lakes such as Lake Washington near Seattle (Edmondson, 1974). However, phosphorus can be easily exchanged between the topmost sediments and the lake water, and this mobility makes phosphorus a somewhat less useful trophic indicator than it might otherwise be (Engstrom and Wright, 1984).

A third measure comes from biological indicators of trophic status. Most aquatic organisms are sensitive to nutrient availability and some, although by no means all, are subsequently preserved in lake-bottom muds. Amongst the most useful indicators are chironomid larvae (Deevey, 1942; Hofmann, 1986; Sadler and Jones, 1997), cladoceran faunas (Frey, 1986) and diatoms (see Technical Box IX). With eutrophication, planktonic organisms typically become more abundant and benthic (bottom-dwelling) forms less so, as the hypolimnion becomes de-oxygenated. As with most palaeoecological studies a combination of different physical, chemical and biological analyses is more powerful than any one technique on its own. These palaeoecological indicators in any case need to be complemented by dating methods that can provide a timescale for environmental changes recorded in the lake sediment profile.

Technical Box IX: Diatom analysis

Diatoms (Bacillariophyceae) are unicellular algae with a shell (or frustule) made of silica. The frustule comprises two, usually symmetrical, valves which are finely sculptured and ornamented with rows of tiny holes (see plate 7.7). Diatom taxonomy is based on the shape, size and ornamentation of the frustule. Although there are several other types of algae, diatoms are one of the few to be preserved in any detail after death. The frustule of all except a few fragile species survives well in sediments and allows taxonomic identification to species or even sub-species level. Diatoms are normally between 5 and 500 microns long, so identification and counting requires a high-powered light microscope and normally takes place at magnifications of 200 ×–1000 ×. In this, and in preparation of slides, analysis of diatoms is similar to that of pollen, except for the different chemical agents employed. Indeed, diatom analysis may reasonably be considered to fulfil the same function for aquatic ecosystems that palynology does for terrestrial vegetation. Compared to pollen, diatoms have two advantages; first, the higher taxonomic resolution possible, and second, the fact that diatom fossils represent the organism itself rather than a part of the reproductive process, with all the biases that this avoids.

Diatoms occupy an extremely diverse set of habitats, including not only freshwater lakes and rivers but also saline waters such as closed lakes, estuaries and the ocean itself, and they are even found in peat and soil. Whilst being an important component of aquatic and marine food chains,

TECHNICAL BOX

Plate 7.7 Diatom frustule of the small centric species
Cyclostephanos tholiformis (*c.*10 microns diameter)

their main value in environmental studies lies in their ecological sensitivity. Using modern species and community
ecology it is possible from fossil diatom assemblages to
reconstruct past pH, salinity, water depth, productivity and
other conditions (Gasse, 1987; Dixit et al., 1992; Reid et al.,
1995). Increasingly, this is undertaken by using numerical
methods, involving a statistical comparison of 'fossil' assemblages with data calibration sets from modern water
bodies (e.g. Birks et al., 1990). Comprehensive reviews of
diatom analysis are provided by Battarbee (1986) and Moser
et al. (1996), and a simple taxonomic key for aquatic forms
can be found in Barber and Haworth (1981).

LOUGH NEAGH

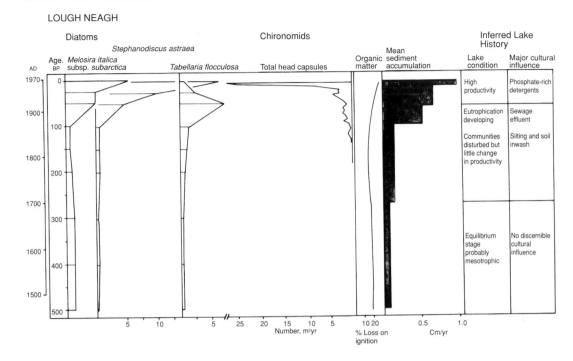

Figure 7.5 Lake sediment record of eutrophication, Lough Neagh, Ireland (based on Battarbee, 1978; Carter, 1977)

[14]C cannot be used for materials younger than 200 years, but other radiometric techniques have now been developed using shorter-lived isotopes and these effectively plug the time gap for recent sediments (see Technical Box X).

The largest lake in Ireland is Lough Neagh, today of markedly eutrophic character. Its present condition is thought to be linked to the input of pollution by domestic and industrial effluent, although the lake's catchment has a long history of agricultural land use which may also have contributed to nutrient enrichment. In order to establish the origins of Lough Neagh's eutrophic state, a number of palaeolimnological investigations have been undertaken (Battarbee, 1978; Carter, 1977). Organic matter does increase somewhat towards the top of sediment cores obtained from the lake (see figure 7.5), but more convincing evidence of trophic change comes from biological indicators, notably diatoms and chironomids. Both organisms show significant changes in species composition in the top metre of the lake sediment profiles, indicating that the present-day lake biota is not a long-term, stable one. Richard Battarbee attempted to go one step further and reconstruct the palaeoproductivity of Lough Neagh through absolute counting of core diatoms. Diatoms, of course, form part of the total algal population and their numbers would be expected to increase with eutrophication. Good dating control is essential in any study of 'absolute' microfossil numbers, and in this case it was provided through the isotopes ^{14}C, ^{210}Pb and [137]Cs. This chro-

nology in fact showed there to have been a marked acceleration in the rate of lake sedimentation after about AD 1700.

The results of this absolute analysis show a dramatic increase in diatom influx during the late nineteenth and twentieth
centuries, with values ten times higher between 1960 and 1970
than they were during the stable base-line state before AD 1700.
There are a number of factors that could prevent these diatom
influx figures from being a reliable measure of former lake productivity. These include problems of diatom frustule dissolu-
tion, redistribution within the lake and loss down the outflow
river, as well as the fact that many other organisms contribute
to primary productivity that are not preserved in the lake
sediment record. Fortunately, the diatom data are given strong
support by a comparable upcore increase in the number of
chironomid head capsules (Carter, 1977). It would therefore
seem that in this case, diatom influx figures do provide a
reasonable approximation to the palaeoproductivity of Lough
Neagh.

What, then, are the most likely causes of Lough Neagh's
enrichment? Archival research has shown that the onset of
eutrophication coincided with the Victorian introduction of
piped sewerage systems (Patrick, 1988). Instead of being used
as land fertilizer and absorbed by the soil system, sewage
was discharged as effluent into rivers and thence into the lake
itself. Phosphorus inputs have continued to increase during
the twentieth century, especially from around AD 1915 when
Lough Neagh came to be used for waste disposal by peripheral
towns, and from the 1950s with the advent of phosphate
detergents. The fact that these cultural changes coincided with
ecological ones in the lake does not constitute proof that the
former caused the latter, but it does represent convincing cir-
cumstantial evidence.

Lough Neagh's present-day ecology is the result of cultural
eutrophication since the Industrial Revolution. This means that
if remedial measures were taken to reduce pollution there is
every prospect of the lake returning to its former less eutrophic
state. The ability to recover following disturbance – or elasticity – is one of the features of lakes that makes them relatively
resilient ecosystems. Once their supply is cut off, pollutants
usually become buried in the sediments at the lake bottom
or are flushed away down the outflow. This has happened in
Lake Washington, where after public pressure, nutrient-rich
waste water was diverted away from the lake after 1967. The
culturally eutrophied lake is now recovering and recent cores
show a declining bulge in the phosphorus content of the lake
sediment column (Edmondson, 1974).

Studies of lacustrine and coastal sediments have shown
similar stories of cultural eutrophication through Europe and

Table 7.3 Diatom- inferred changes in total phosphorus (TP) for selected European lakes		Minimum		Maximum	
		TP (µg l⁻¹)	*Date*	*TP (µg l⁻¹)*	*Date*
Augher, Northern Ireland		35	pre-1900	130	1968
Væng Sø, Denmark		100	1900	160	1987
Langesø, Denmark		135	pre-1900	225	1975
Marsworth, England		180	1900	486	1983

Source: data from Anderson et al. (1993), Anderson and Odgaard (1994), Bennion (1994)

North America (e.g. Anderson, 1989; Cooper and Brush, 1991; O'Sullivan, 1992; Turner and Rabalais, 1994). Numerical calibration of these data, for example using weighted averaging techniques, now permits precise quantitative reconstructions to be made of past phosphorus and nitrogen levels (table 7.3). Anomalously productive ecosystems in regions of otherwise oligotrophic lakes, such as Shagawa Lake, Minnesota, have been revealed as the product of human impact, in this case from mining and sewage waste disposal (Bradbury and Waddington, 1973). The cultural causes of eutrophication are activities associated with industrial and urban society – piped sewage, chemical fertilizers and detergents, and industrial waste. In non-industrialized regions of the world, pollution by nutrient-rich waste water has been a much less significant problem in the past, although this is changing fast with Third World urbanization. Eutrophic lakes do exist in the tropics, but not always as a result of pollution. Lake George in Uganda, for example, is a highly productive ecosystem with dense crops of blue-green algae and diatoms, but palaeolimnological studies have shown it to have been that way for at least the last 500 years (Viner and Smith, 1973; Haworth, 1977). An attempt to 'rectify' this naturally eutrophic lake would be unlikely to succeed!

Acidification and atmospheric pollution

If eutrophication came to be recognized as a major environmental problem during the 1960s and 1970s, its equivalent *par excellence* in the 1980s was acidification. Increased atmospheric deposition of acids is believed to have wide-ranging environmental consequences, including tree death, accelerated weathering of building facades and damage to human health. However, probably its most widespread impact, and that which is discussed here in detail, is on freshwater ecosystems. In many respects, lake and stream acidification represents the opposite trend to that of eutrophication. Whereas eutrophied waters are murky but nutrient-rich and productive, acidified ones are clear but nutrient-poor and infertile. It has been proposed that acidification starts with the emission of oxides of sulphur (SO_2) and nitrogen (NO_x) into the air. These are derived from car exhausts and, more importantly, from coal-fired power stations. It has long been known that atmospheric pollution in urban areas leads to 'dry' fall-out of sulphur, and that this poses an ecological threat, for example to lichens. Precisely in order to avoid the problem of local dry fall-out of pollutants, high stacks were constructed from which emissions would be carried up and away into the atmosphere. It was hoped that the resulting dilution would be the solution to the pollution!

Once up in the atmosphere SO_2 and NO_x undergo chemical transformation, involving combination with water vapour to produce dilute sulphuric and nitric acids. This 'acid rain' is carried downwind by atmospheric circulation and is returned to Earth as 'wet' deposition in the form of rain or snow. Hydrological processes then lead to the acidified waters entering streams and lakes, with consequences for their plant and animal communities. Some organisms thrive in a strongly acidic environment, but most do not, and there is normally a decline in both total biological activity and in species diversity below pH $c.6.0$. In the case of 115 southern Swedish lakes algal species numbers were found to decrease dramatically below a threshold level of pH 5.8 (Almer et al., 1974). The precise ecological mechanisms associated with acidification of freshwaters are only partially understood, but one important aspect is the mobilization at low pH of aluminium, which is toxic to many higher organisms such as fish. A decline in fish stocks is one of the main characteristics of acidified waters, and one that has caused Scandinavians and Canadians much anxiety. What makes them not just anxious but angry is that the pollution appears to have been exported to them by their neighbours in central and western Europe and the United States. Acidification is an international problem in which the perpetrator might hope to escape notice by being hundreds of kilometres from the scene – or at least the consequences – of the crime.

The account just presented is, not surprisingly, a greatly simplified and stylized scenario. For example, one lake watershed may be experiencing acidification whilst another immediately adjacent shows no such sign. Acid precipitation clearly does not have the blanket effect over the landscape that might be expected from atmospheric deposition; catchment conditions must also be an integral part of the problem. In particular, the ability of catchment soils and bedrock to neutralize acid precipitation will critically influence the pH of the resulting runoff. Buffering capacities will be high in areas of basic bedrock such as limestone, and low on granite and other acidic

TECHNICAL BOX

Technical Box X: Dating by short-lived radioisotopes (^{210}Pb and ^{137}Cs)

Lead-210 (^{210}Pb) is an unstable isotope with a half-life of just over 22 years and a dating range of 100–200 years, thus conveniently complementing ^{14}C (see chapter 2). Unlike ^{14}C, on the other hand, ^{210}Pb forms part of a longer decay chain which, in effect, begins when radium-226 (^{226}Ra) is released from bedrock as part of catchment erosion. Some will enter streams and lakes and will decay *in situ* to form ^{210}Pb. However, that is not the only source of this isotope. A proportion of ^{226}Ra escapes into the atmosphere as the inert gas radon-222, which decays via a series of short-lived isotopes to form ^{210}Pb. This excess of unsupported ^{210}Pb returns to Earth as rainfall or as dry fall-out to augment that produced by *in situ* decay (see figure 7.6). Only the unsupported ^{210}Pb is required for dating purposes. Dates are derived in a similar way to that for radiocarbon, with a down-profile decline in unsupported ^{210}Pb in accordance with natural radioactive decay of the isotope. Various models of age estimation exist, the most comprehensive of which assumes a constant rate of supply over time (Appleby and Oldfield, 1978). In practice, however, changes in lake hydrology or catchment soil erosion rate can result in an irregular flux of unsupported ^{210}Pb (Oldfield and Appleby, 1984).

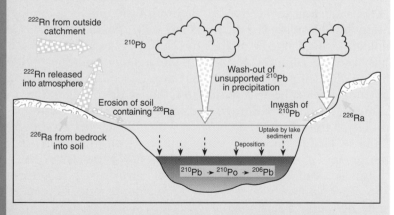

Figure 7.6   Lead-210 pathways in a lake catchment (based on Wise, 1980)

^{210}Pb and ^{14}C are naturally occurring radioisotopes, but there also exist others of artificial origin, one of which is caesium-137 (^{137}Cs). The main source of these artificial isotopes has been the atmospheric testing of nuclear weapons, which started in the 1950s, reached a maximum in the late 1950s and early 1960s, and subsequently declined with the advent of the nuclear test ban treaties until the Chernobyl nuclear accident in 1986. ^{137}Cs began to be deposited in 1954, and fall-out peaked in the northern hemisphere around 1963 (see figure 7.7). This and the later Chernobyl peak can be distinguished by also measuring another caesium isotope, ^{134}Cs, which was released in the Ukrainian accident. Investigations have shown that ^{137}Cs is strongly adsorbed onto sediment or soil, and that it has been incorporated into lake sediments deposited in the last 30–40 years. The changing concentration of ^{137}Cs in the uppermost sediment layers of lakes in eastern North America and the English Lake District has been shown to parallel the recorded deposition of the isotope on land (Pennington et al., 1973). ^{137}Cs thus

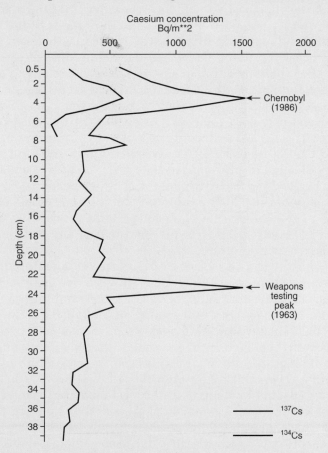

Figure 7.7 Stratigraphic deposition of radioactive caesium at Ponsonby Tarn in the English Lake District (modified from Bonnett and Cambray, 1991)

acts as an effective tracer suitable for studying soil erosion and sediment accumulation rates during recent decades (Wise, 1980; Walling and Quine, 1990). ^{137}Cs also provides a useful cross-check for ^{210}Pb chronologies, and a combination of the two methods provides a strong basis for determining the timing of recent human impact on ecosystems.

rocks. Vulnerability to acidification will therefore be greatest where there is a combination of high acid loadings and low catchment buffering capacities. Consequently, any reduction in emissions will be most effective if it is targeted at vulnerable sites rather than applied 'across the board'. For this, a **critical loads** concept is widely utilized.

Amongst the most vulnerable areas where the critical load is most quickly exceeded are the southern portions of the Canadian Shield and Scandinavia, which once lay beneath major Pleistocene ice sheets. These areas have thin soils but innumerable lakes, most of which in any case have had relatively acidic waters. But how acidic? Some researchers have suggested that progressive natural acidification during the course of the Holocene may go a long way to explaining the existence of presently acid lakes. Records of water quality in the English Lake District from the 1940s onwards indeed show little sign of the decline in pH over recent years (Sutcliffe et al., 1982). Yet chemical and fishery records in southern Scandinavia show a more systematic downward trend over the same period (Almer et al., 1974; Wright, 1977). A longer-term perspective would seem necessary to resolve this discrepancy and test the progressive acidification hypothesis.

pH history can be measured from lake sediment profiles by analysing chemical composition, or from the hard parts of biological indicators such as invertebrates and chrysophytes (Smol, 1986). As with eutrophication, one of the most effective palaeoacidity gauges comes from diatom analysis. Diatoms were classified by the German diatomist, Hustedt, into five different pH groups, ranging from species found only in alkaline waters (alkalibiotic) to those found only in acid and not in alkaline or neutral waters (acidobiontic). Working in southern Sweden, Ingemar Renberg and Tor Hellberg (1982) used Hustedt's scheme to produce a numerical index (ß) by which pH could be reconstructed from diatom assemblages. In 30 modern lake samples, the log of index ß had a correlation with water pH of –0.91. Renberg and Hellberg then applied this index to diatom core samples from a number of presently acid Swedish lakes. At Gårdsjön, they calculated that there had been a decline of *c*.1.0 pH unit during the early part of the Holocene. This was a natural change brought about by progressive leaching of catchment soils which had been base-rich at the start of the post-glacial, but which later became acid and peaty. Lake-water acidity then stabilized at just below pH 6.0, that is until the top few centimetres of the sediment core. Here there was an abrupt fall of over 1.5 pH units, more than had occurred during the whole of the previous 10 000 years. On the basis of the abundance of oilsoot particles in the lake sediment, Renberg and Hellberg dated the onset of recent acidification to about AD 1950. Data from Gårdsjön and other acidic northern European lakes (e.g. Jones et al., 1989; Renberg, 1990) have shown that they are definitely not in their 'natural' state – their base-line equilibrium has been seriously disturbed during the last 100 years or so. What is more, this disturbance has taken lakes below a key ecological threshold at around pH 5.5–5.8. More recent reconstructions of lake acidification have used the pH optima of individual species calibrated by a weighted averaging technique (Birks et al., 1990). This is statistically sounder than the index ß approach, but the results are essentially similar.

Although the progressive acidification hypothesis could now be ruled out as the sole explanation, this did not make acid precipitation necessarily guilty either. Closer examination was required to establish exactly which human activities have been responsible for recent ecological impact on acid freshwaters. Apart from acid precipitation, two other possible anthropogenic causes have been proposed, both related to land-use changes in lake watersheds. The first is afforestation, particularly involving conifers, which can cause increased uptake of nutrients and thereby reduce soil buffering, and can also trap and concentrate acid gases. The second land-use change involves the decline of traditional upland farming based on sheep grazing and regular woodland burning (Rosenquist, 1978). The resulting regeneration of acid heathland plant communities, it has been argued, is also capable of releasing sufficient soil acids to produce freshwater acidification.

In the Galloway lakes of southwest Scotland, Richard Battarbee, Roger Flower and colleagues from University College, London set out to test these competing cultural hypotheses. They did this by trying to eliminate – in true 'whodunnit?' style – the possible culprits (Battarbee et al., 1985). Their first suspect was afforestation, which they considered by taking four lake catchments, two of which had been recently coniferized and two of which were unforested. Diatom analysis of lake cores, calibrated statistically, showed essentially similar sequences; all four lakes had been acidified by about 1.0 pH unit from their original base-line condition, much as at Gårdsjön

Table 7.4 Acidification data for Galloway lakes, Scotland, based on diatom calibration and ^{210}Pb dating

	Pre-acidification pH	Modern pH (predicted)	Modern pH (observed)	pH decline	Date of tree planting	Date of onset of acidification
Loch Enoch	5.2	4.3	4.4–4.7	0.9	–	1840
Round Loch of Glenhead	5.7	4.7	4.5–5.0	1.0	–	1850
Loch Grannoch	5.6	4.4	4.4–4.9	1.2	1962, 1976	1925
Loch Dee	6.1	5.6	4.9–5.9	0.5	1976	1890

Source: Battarbee (1984)

(Flower and Battarbee, 1983; Kreiser et al., 1990). Furthermore, ^{210}Pb dating showed acidification to have begun in the late nineteenth and early twentieth centuries (see table 7.4), 50 years before afforestation began. The planting of conifers may have aggravated the problem, but certainly did not start it.

The second land-use hypothesis was somewhat harder to test in the absence of an adjacent heathland site with a low rate of acid deposition for comparison. However, pollen profiles spanning recent centuries from Loch Enoch in Galloway showed no increase in ling (*Calluna vulgaris*) as would be expected with heathland regeneration, but instead a decline relative to gramineae pollen (Battarbee et al., 1989). Nor did historical records indicate wholesale land abandonment or decline in grazing, with the inference that this explanation too could be rejected. The only suspect remaining was acid precipitation, a conclusion also reached by comparable research in eastern North America (Charles et al., 1987; Davis, 1987).

Many other atmospheric pollutants have also been recorded stratigraphically, in environments ranging from peat profiles through alpine lake sediments to ice cores (e.g. Oldfield et al., 1981; Chambers et al., 1979; Renberg et al., 1994). Metals such as lead, zinc (Zn) and copper (Cu) are found at higher levels in acidified lakes, and they have typically increased as pH has fallen (figure 7.8). While it is unlikely that these trace metals, or the fly ash particles found in lake cores, have themselves significantly contributed to reduced pH, they do give strong support to the link between industrial pollution of the atmosphere and changes in freshwater ecosystems (Battarbee, 1994). In the Baltic republic of Estonia atmospheric pollution has not caused acidification, but the opposite! Northeast Estonia contains the world's largest commercially exploited oil shale deposit, and combustion of it in major thermal power plants has led to the release of calcium-rich fly ash which has a pH value

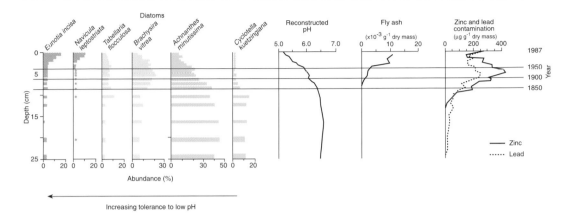

Figure 7.8 pH history of Loch Chon in the Trossachs of central Scotland reconstructed numerically from diatom taxa, along with the record of atmospheric deposition of fly ash, zinc and lead (after Kreiser et al., 1990)

as high as 12! Much of the surrounding region comprises lakes and peat bogs that are naturally acidic, and whose ecology is threatened by this alkaline air-borne pollutant. Analysis of peat and lake cores has shown that since the 1950s, acid-loving taxa have been replaced by those preferring alkaline waters, amounting to an increase of more than 1 pH unit (Punning, 1994). *Sphagnum* peat has been destroyed and about one-third of the typical bog vegetation has disappeared.

Public pressure, backed by this and related scientific research, has encouraged some governments and industries to reduce pollution emissions since the 1970s. But is this 'clean-up' bringing a tangible improvement in environmental quality in wilderness areas? On the remote ice cap of Greenland, atmospheric deposition of lead (Pb) increased 20 times between *c*.1800 and 1970 (Murozumi et al., 1969). However, lead concentration in Greenland snow profiles has *decreased* substantially during the 1970s and 1980s, in response to the decline in use in leaded petrol, particularly in the United States (Rosman et al., 1993). Similarly, mercury levels in lake sediments, which had increased fourfold in the US Mid-West since AD 1800, have shown a decline since 1980 (Engstrom and Swain, 1997). Monitoring of acidified lakes has also shown some pH values starting to rise again after decades of decline. Will these acidified lakes be able to recover biologically, and, if so, how long will it take? An encouraging illustration of the powers of recovery (or elasticity) of lake ecosystems comes from the Sudbury area of Canada, which has one of the world's largest smelting plants and tallest chimney stack. Palaeolimnological investigation of 22 out of more than 7000 lakes in the immediate damage zone around Sudbury had shown a twentieth-century decline in lake pH and increases in aluminium and zinc deposition (Dixit et al., 1989). However, reverse acidification has started to occur in some sites as sulphur emissions have been regulated (figure 7.9). Chrysophyte

Figure 7.9 Reverse acidification recorded in the sediment profile of Swan Lake, Canada (after Dixit et al., 1989). Right-hand graph shows SO₂ emissions from the Sudbury stack

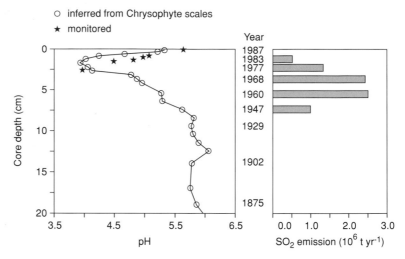

scales in a short sediment core from Swan Lake also show a reversal of the previous trend, with a time lag of only a few years for biological recovery (Dixit et al., 1989). A comparable biological reversal in diatom floras has been detected in some Scottish lochs as acid deposition has started to decrease (Allott et al., 1992). Higher organisms like brown trout (*Salmo trutta*) and the natterjack toad (*Bufo calamita*) will doubtless take longer to return than this. None the less, it offers persuasive evidence that pollution controls can often soon be followed by ecological improvement.

It is well known that the advent of industrial capitalism caused the input of both atmospheric and terrestrial pollutants into the environment to shoot upwards. What palaeoecological records have been able to show is first, the widespread nature of their deposition, second, how previously stable aquatic eco-systems have been seriously disturbed in consequence, but third, how many of these systems are self-cleansing once pollutants are reduced below critical load thresholds.

CHAPTER EIGHT

The Environmental Future: A Holocene Perspective

The history of mankind is a long and diverse series of steps by which he has achieved ecologic dominance . . . largely he has prospered by disturbing the natural order.

C. Sauer (1952)

Some of those steps have been described in preceding chapters of this book and are summarized in figure 8.1. During the late Pleistocene natural agencies – particularly climatic change – played the dominant role in determining the character of environmental systems. The glacial-to-interglacial transition brought dramatic changes to environmental systems, with sea levels rising and biota striving to keep pace with rapid shifts in climate. The start of the present interglacial was consequently a period of dynamic change and landscape instability, marked in many areas by high rates of erosion and sediment flux (Dearing, 1994). By the early Holocene, the world's environments and biomes had begun to take on their modern 'natural' form, although this process was not complete until the mid-Holocene in the Arctic and the tropics. The opportunities offered by the Holocene world in turn encouraged new forms of cultural adapation to emerge, including the domestication of plants and animals.

It can be observed too how relations between nature and culture have been transformed as different modes of production have emerged or declined over the course of the Holocene. Early human–environment relations, for instance, involved a direct dependence on wild plant and animal resources. Culture was continuous with nature, as it is in contemporary h-f-g groups such as the Ba-mbuti pygmies, for whom the tropical rainforest is a mother and a home, not a wilderness to be tamed. With agriculture came greater cultural control over the relationship with nature, and the widespread creation of semi-natural agro-ecosystems. Animals were no longer blood relations but servants. But if peasant farmers and nature formed a separate dualism, they none the less constantly interacted to shape each other dialectically. And while hunter-gatherers and peasants are in some respects opposites, they share in common an intimate relationship with their natural habitat and a deep respect for nature. Only under contemporary industrial society has that interdependence been lost and the separation of people and nature been made complete (see figure 8.2). The relationship between industrial capitalism and the natural environment is well illustrated by the example of acid freshwaters, whose history has been so effectively revealed by palaeolimnology (Battarbee, 1984; Davis, 1987). Even though the industrially based processes of wealth creation are physically remote from boreal forest lakes, they have been capable – via atmospheric pollution – of destroying the stability of these ecosystems through acidification.

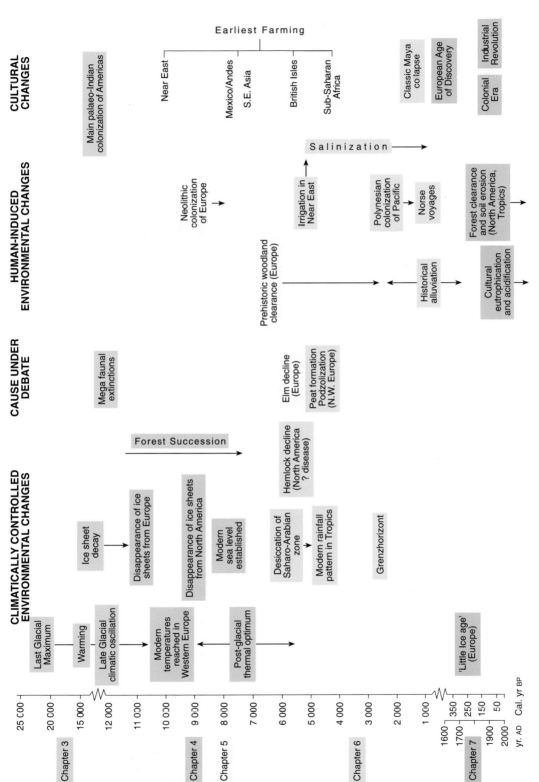

Figure 8.1 Summary chart of Holocene environmental and cultural changes

Figure 8.2 The changing relationship between humans (H) and the natural environment (E) over the course of the Holocene, including **A** nature of interaction and **B** relative impact

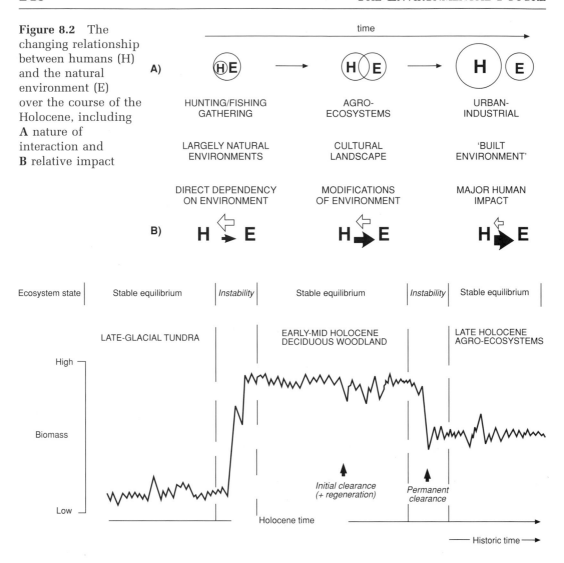

Figure 8.3 Metastable equilibrium model of Holocene ecological change in northwest Europe

When viewed over the long term and the large scale, human impact upon the natural world seems to have increased relentlessly. But in reality this seemingly inexorable increase in impact has been formed out of an agglomeration of specific regional histories, each one different. Contrast, for example, the histories of human–environmental relations in the Near East, where the impact of agriculture was felt even before the last glaciation had ended, and in the island ecosystems of the eastern Pacific, which were completely free of human disturbance until the Polynesian arrival less than 2000 years ago. At regional scales such as these, environmental histories often display marked discontinuities, and at local scales this is even more true. In other words, at local and regional scales it may be more appropriate to see the changing face of nature as a

Table 8.1 Factors determining ecosystem response to disturbance

Ecosystem resistance	robust --------------------------------- (e.g. grassland)	fragile (e.g. island ecosystems)
Ecosystem recovery	elastic --------------------------------- (e.g. lakes)	irreversible (e.g. badland erosion)
Origin of disturbance	internal/gradual -------------------- (e.g. soil leaching)	external/abrupt (e.g. new human arrivals)
Duration of disturbance	short-lived --------------------------- (e.g. pathogen)	sustained (e.g. marine incursion)
Continuity of disturbance	episodic ----------------------------- (e.g. fire events)	progressive (e.g. eutrophication)
Overall ecosystem response	maintenance of stability/recovery	instability/degradation/collapse

series of metastable equilibrium states than as following a progressively changing trajectory (Oldfield, 1983). According to the metastable equilibrium model, prolonged periods of environmental stability were punctuated and separated by much shorter-lived phases of rapid, sometimes catastrophic, change (see figure 8.3). These unstable events represent ecological crises, during which key thresholds were crossed.

Holocene Environmental Crises

While palaeoecological records have helped link contemporary environmental problems of pollution and land degradation with the emergence of industrial capitalism and colonialism, they also show that environmental crises are by no means the monopoly of modern times. Human-induced ecological problems have occurred throughout the time period since the last ice sheets started to wane over Europe and North America, and have involved societies operating under very different modes of production. The catastrophic extinction of most of America's large mammal species around 11 000 years ago would almost certainly not have occurred without the involvement of palaeo-indian hunters (Martin and Klein, 1984). The formation of blanket peat bogs over much of western Britain and Ireland during the mid-Holocene was triggered in many cases by the felling of trees by prehistoric pastoralists and farmers (Moore, 1975). The extinction of the moa and other flightless birds in New Zealand and the deforestation of Easter Island were both directly consequential upon the arrival of the Polynesian peoples 1000 years ago (Flenley and King, 1984; McGlone et al., 1994).

Nor are environmental crises uniquely the responsibility of human beings. The catastrophic demise of the North American spruce forest at the start of the Holocene occurred as it was squeezed against the retreating Laurentide ice sheet (Watts, 1983). Other biotic crises, such as the decline of hemlock in eastern North America 4800 years ago, were probably caused by pathogen (disease) attack (Allison et al., 1986). The late Pleistocene and Holocene are also notable for the abrupt shifts in climate (Lamb et al., 1995). Climatic crises, in the form of droughts several centuries long, have occurred in the tropics every 2000–3000 years during the Holocene (Gasse and van Campo, 1994). Abrupt climatic changes are also recorded around the North Atlantic, where beetle remains and ice-core records show that mean temperatures rose dramatically on two occasions during the glacial-to-interglacial transition (Atkinson et al., 1987; Alley et al., 1993). Not surprisingly, abrupt climatic events had a marked effect on the stability of some environmental systems and human societies (e.g. Weiss et al., 1993).

What circumstances produced these environmental crises, and what lessons can be learned from them to help avert them in future? The main relevant factors, listed in table 8.1, may be condensed down to ecosystem resilience on the one hand, and the type of impact on the other. Some ecosystems are robust, like grassland environments, or are 'elastic' in their recovery, like lakes. However, others, such as the limestone plateau of the Burren in western Ireland, had a low threshold for change coupled with an inability to recover following disturbance. Whether the changes in a particular environmental system are reversible or not frequently depends on the maintenance of a soil suitable for plant growth. In the case of the Burren, the soil erosion that followed prehistoric and early historic forest clearance was so severe that the landscape was left as ecologically degraded bare limestone (Drew, 1982; Watts, 1984). It therefore appears that the preservation of the soil is a vital factor in the maintenance of ecological habitats as a whole. This is one reason why soil degradation has been and still is such a critical environmental issue, be it in the Burren 3000 years ago, Oklahoma in the 1930s or in Lesotho today.

The Burren needed only a small 'push' for it to cross the threshold into environmental instability, and the same is true of the history of other inherently fragile ecosystems like those parts of upland Britain today blanketed by peat bog. With such ecological shifts as peat initiation, it is wrong to talk of human and natural causation separately, for human impact would have been ineffective without suitable environmental preconditions. In fact, most Holocene ecological crises have involved several factors operating in combination. The European elm decline, for instance, is now most plausibly explained by the combined impact of Neolithic farming and a pathogen outbreak (Peglar and Birks, 1993). Whereas the impact of either factor on its own might have been absorbed by the elm tree population, the combination of both factors produced environmental stresses that could not be withstood. Significantly, in North America the hemlock did eventually recover to a status similar to that before its decline, while in Europe the elm did not. If there is a moral here for contemporary environmental problems, it is that human disturbance may not on its own produce ecosystem breakdown, but that it will weaken the resistance of ecosystems to other environmental impacts.

Another very important influence on Holocene environmental crises lies in the origin of disturbance. Abrupt, external impacts have usually had much more dramatic consequences than gradual, internally generated changes. This applies to disturbances of natural origin – for example, abrupt shifts in climate – just as it does to cultural ones, such as the arrival of the Polynesians in New Zealand. The degree of abruptness goes a long way towards explaining why late Pleistocene mammalian extinctions were so severe in the Americas but were hardly felt at all in Africa. Hominids had co-existed with the megafauna of the African savannas for millions of years, and the two had evolved side by side and become adapted to one other. By contrast, *Homo sapiens* only arrived in the New World at the end of the Pleistocene, expanded rapidly and was unfamiliar to the indigenous animal population. In other words, human disturbance increased gradually in Africa, giving the biota time to adjust, whereas in the Americas the impact came abruptly and catastrophically.

Environmental Conservation and Holocene History

Palaeoenvironmental studies help in understanding the trajectory which a particular environmental system is following and hence aid in its management. They can establish, for example, whether present rates of soil loss are typical of the Holocene or whether they have been accelerated in recent times. They can further help in the identification and designation of ecosystems that have survived essentially undisturbed through the Holocene. For example, Lake Baikal in Siberia is the world's largest and deepest freshwater lake and has a rich and unique biota, including a freshwater seal. Analysis of short cores from the lake has shown that while the uppermost sediments have begun to be contaminated with pollutants like lead, there has

so far been no detectable ecological effect of human disturbance (Flower et al., 1995). In other, already polluted lake
ecosystems, records stored in their bottom muds allow their
pre-disturbance base-line condition to be identified, which can
then serve as a restoration management target.

In undisturbed environments, such as ancient woodland,
genetic and other characteristics are retained which are absent
in more disturbed areas of secondary woodland. Preservation
of the gene pool is especially important where little of the
original vegetation remains intact. For instance, only 0.2 per
cent of Ireland lies under 'native' woodland – equivalent to
the area occupied by the city of Dublin – and most of this is
secondary. Pollen studies have shown several western Irish
'ancient woods' in fact to be of recent origin, and indeed that
they are still evolving in terms of their species composition
(Hannon and Bradshaw, 1988; Mitchell, 1990). The recognition that fire is a long-standing agent in many forests is another
way in which ecological history has helped in environmental management. Fires can act to release and make available
mineral nutrients, regulate dry matter accumulation and control forest insects and parasites (Wright and Heinselman, 1973).
Misguided fire suppression by forest managers has sometimes
had serious adverse effects on forest ecology and increased the
risk of a major, uncontrollable conflagration. Moreover, fire scars
on dated tree-rings and charcoal peaks in lake sediments can
help to establish the 'natural' magnitude and frequency of
forest fires during the period before written observations, and
before twentieth-century fire suppression policies were applied
(Millspaugh and Whitlock, 1995; Larsen, 1996).

The example of fire ecology illustrates how far removed from
nature urban, industrial society has become in its philosophy
of environmental conservation. According to this philosophy,
the natural environment should be preserved as it is rather
than being allowed to change dynamically. However, very little
in the natural world is stable when viewed over the timescale
of centuries and millennia, and nature should not be made
sacred in the name of science. Individual species have risen
and fallen in importance during the Holocene, making the
most of opportunities presented to them. For example, beech
was relatively unimportant in the British Isles until agricultural clearance provided gaps in the forest for it to expand late
in the Holocene, and it was absent in Ireland until the last
millennium (Godwin, 1975). By contrast, pine was common in
Ireland for most of the Holocene, but died out 1000–2000 years
ago (Bradshaw and Browne, 1987). Trying to use species
conservation to stop the ecological clock at AD 2000 would be
as artificial as designating one of these trees to be a native
species but the other not.

The national park idea, developed in North America and subsequently exported to other parts of the world, reflects the philosophy that the natural environment should be preserved as sacrosanct rather than utilized. The International Union for the Conservation of Nature and Natural Resources (IUCN) describes a national park as 'one or several ecosystems not materially altered by human exploitation or occupation . . . and where the highest authority has taken steps to prevent or eliminate as soon as possible exploitation or occupation in the area' (IUCN, 1975).

In reality many ecosystems are far from being wholly 'natural', and instead owe their distinctive character to particular patterns of land use or other human actions. Garaet el Ichkeul in Tunisia is a shallow coastal lake which is one of the most important wetland sites for migratory wildfowl around the Mediterranean, but its present ecological state only began in the nineteenth century when French engineers built a ship canal into an adjoining lake and inadvertently allowed the inflow of saltwater (Stevenson and Battarbee, 1991). Without this unintended human impact, Ichkeul would not have its present high conservation value! Similarly, many of our most valued 'natural' landscapes – such as the wilderness of Dartmoor and the scrub oak-pine barrens of New England – are plagioclimax ecosystems created and partly maintained by traditional land-use practices such as grazing and burning. The idea that the wilderness should be free of people living and working there has less to do with successful ecosystem functioning than it has with the romantic ideal of a primaeval nature independent of 'man'. Palaeoecology shows that ideal to be a myth. In all but a few cases humans have been an integral part of ecosystems since present-day climates and environments became established early in the Holocene. For most ecosystems it is therefore effectively impossible to study environmental history separate from cultural history, and vice versa.

The 'best leave alone' conservation philosophy makes even less sense in those parts of the world – often lumped together as the Third World – where peasant farming, pastoralism or h-f-g today remain strongly embedded. Societies operating under these modes of production have, over time, slowly but progressively established a detailed knowledge of their local habitats. This indigenous environmental knowledge, for example of tree species and their different uses, is geographically specific and cannot be built up overnight. In these cases, conservation needs to include cultural traditions as much as it does flora and fauna.

But the real issue is not one of survival and preservation but of incorporating traditional cultural ecologies into future development. As Paul Richards (1985) has demonstrated for

West Africa, indigenous peasant farmers had a far better understanding of ecologically appropriate forms of production (e.g. swidden) than outside specialists such as colonial agricultural officers. Building on such traditional forms of culture–environment relations from the bottom up is more likely to be successful than imposed, alien technological 'solutions' (Redclift, 1984). A good example is provided by wetland horticultural systems, which have a long antiquity in the tropics (e.g. Golson, 1977; Darch, 1988). What is more, palaeoenvironmental research has shown these to be ecologically sustainable, nutrient-conserving systems which do not normally cause long-term soil degradation. Small-scale garden horticulture remains important in many parts of the tropics and offers an alternative to conventional large-scale irrigation schemes (Lambert et al., 1990).

Perhaps the most pressing of the world's contemporary conservation issues is the continuing destruction of tropical forests, especially in Amazonia. Sadly, the present pattern of exploitation shows few signs that the lessons of environmental history have been learned and acted upon (compare with table 8.1). Clear-felling not only pushes the rainforest beyond the threshold of ecological stability, but also leads to severe soil erosion and hence makes forest regeneration difficult or even impossible. It ignores indigenous, traditional forms of land use and instead employs ecologically insensitive, alien agricultural systems. To judge by example from the environmental past, abrupt external impact on a fragile environment has not usually been conducive to ecological stability. But lest this book end pessimistically, let it be noted that environmental history also offers some hope for the future. Human-induced forest clearance has so far been no worse than the naturally caused ice-age deforestation of the Earth, and during the Holocene the forests – tropical and temperate – were able to reclaim most of that area within a thousand years. This was a mere speck of time for the Earth's environmental history, but can we afford to wait that long?

APPENDIX: CALIBRATION TABLE FOR RADIOCARBON AGES

^{14}C age (yr BP)	Calendar age (Cal. yr BP) intercepts and range	Calendar age (Cal. yr BP)	^{14}C age equivalent (uncalibrated yr BP)
500	519	500	430
1 000	928	1 000	1 120
1 500	1 354	1 500	1 600
2 000	1 938	2 000	2 060
2 500	(2 709) − 2 581 − (2 509)	2 500	2 450
3 000	(3 205) − 3 187 − (3 169)	3 000	2 900
3 500	(3 817) − 3 792 − (3 725)	3 500	3 320
4 000	4 436	4 000	3 670
4 500	(5 246) − 5 122 − (5 056)	4 500	4 010
5 000	5 731	5 000	4 430
5 500	6 294	5 500	4 790
6 000	(6 854) − 6 821 − (6 814)	6 000	5 260
6 500	7 384	6 500	5 720
7 000	7 787	7 000	6 130
7 500	(8 313) − 8 220 − (8 218)	7 500	6 700
8 000	(8 950) − 8 784 − (8 767)	8 000	7 260
8 500	(9 480) − 9 474 − (9 456)	8 500	7 760
9 000	9 979	9 000	8 110
9 500	(10 537) − 10 513 − (10 482)	9 500	8 560
10 000	(11 700) − 11 500* − (11 000)	10 000	9 040
11 000	12 917	11 000	9 800
12 000	13 992	12 000	10 230
13 000	15 437	13 000	11 090
14 000	16 792	14 000	12 000
15 000	17 916	15 000	12 740
16 000	18 876	16 000	13 380
17 000	20 105	17 000	14 180
18 000	21 484	18 000	15 070
		19 000	16 120
		20 000	16 920
		21 000	17 630
		22 000	18 410

Source: OxCal. 14L (except for*)

GLOSSARY

These words appear in bold at their first occurrence in the text of the book.

Acidification The process by which soils or freshwaters become more acid, i.e. come to contain a higher concentration of hydrogen ions. This may be natural or cultural in origin.

Allochthonous Materials deriving from, or processes occurring, outside the system, such as a lake; opposite of **autochthonous**.

AMS dating (AMS = **Accelerator Mass Spectrometry**) A method of **radiocarbon** dating involving direct measurement of the ratio of different carbon isotopes. It can be applied to very small samples.

Anaerobic (or **anoxic**) Without oxygen.

Autochthonous Materials produced *in situ*; or processes occurring within a system; opposite of **allochthonous**.

Bio-archaeology The study of plant and animal remains from archaeological sites in order to reconstruct their economy and environment.

Biogenic Materials of organic origin.

Correlation The matching of strata or events of similar age between one place and another; e.g. correlating sediment cores within a lake.

Critical load The highest influx of polluting compounds that will not cause long-term harmful effects on ecosystems.

Dendrochronology Dating by counting the annual growth rings in trees and old timbers.

Diatoms A family of unicellular algae, whose cells possess a finely sculptured outer wall made of silica, which is often preserved in lake or marine sediments after death. Diatom analysis of lake muds forms an important part of **palaeolimnology**.

Domestication The process whereby wild plants are taken into cultivation and wild animals tamed. This usually involves sowing as well as harvesting of plants, tilling of soil by hoe or plough, and controlled breeding of livestock.

Dystrophic Lakes and streams which are themselves infertile, but which contain dissolved humic substances of **allochthonous** origin, giving a characteristic brown colour to the water.

ENSO An abbreviation for the quasi-periodic climatic oscillation known as El Niño–Southern Oscillation. This is related to the strength of the atmospheric pressure gradient across the South Pacific. Every five to ten years it switches from its normal mode, causing heavy rainfall in areas – like the Peruvian desert – which are normally dry. Flood-drought cycles believed to be related to ENSO have been identified in many parts of the tropics and sub-tropics.

Ergodic The idea that parallel situations can occur in space and time, often involving the substitution of space for time, or vice versa. The woodlands of eastern North America encountered by the first European settlers, for example, have sometimes been thought of as resembling the original, pre-agricultural landscapes of north-west Europe.

Eustatic The global component of sea-level change, primarily involving the interchange of water between land ice and the sea. As ice sheets melt so the ocean volume increases and sea levels rise.

Eutrophic Ecosystems such as lakes and soils that are rich in nutrients and have a high primary productivity. In hyper-eutrophic environments species diversity decreases and predatory organisms decline.

Eutrophication The process of nutrient enrichment, which may be natural or cultural in origin.

Feral Domestic plants or animals which have escaped and returned to the wild.

Flandrian The name sometimes given to the Holocene in northern Europe.

Geo-archaeology The study of sediments at, and adjacent to, archaeological sites in order to reconstruct local environmental histories.

Grenzhorizont A pronouced boundary horizon in northwest European raised mires marking a change from dark, humified peat to light, less well-humified peat. It has been interpreted as marking a shift from a drier (sub-Boreal) to a wetter (sub-Atlantic) climate at *c.*2600 Cal. yr BP, although radiocarbon dating has shown the horizon to be **time-transgressive** rather than **synchronous** across Europe.

Holocene The post-glacial epoch, conventionally – if arbitrarily – dated as beginning 10 000 ^{14}C (or about 11 500 calendar) years ago. Holocene, meaning 'wholly recent', was introduced as a term by Gervais in 1869 and accepted by the International Geological Congress in Portugal in 1885.

Hydrosere A succession from open water to mire and bog at the edge of a lake.

Interstadial A warm climatic interval within an otherwise cold stage, of shorter duration than true interglacial, e.g. Windermere interstadial (15–13 000 Cal. yr BP).

Isostatic Changes involving compression loading and unloading of the Earth's crust, most importantly by ice. Glacio-isostatic pressure beneath ice sheets depressed parts of Canada and Scandinavia by over 1 km at glacial maxima. In some coastal areas isostatic movements have had conspicuous effects on local sea-level histories.

Isotopes Different forms of the same chemical element, defined by the number of neutrons present. Unstable isotopes, such as ^{14}C and the various isotopes of uranium and thorium, experience radioactive 'decay' and can be used for dating purposes. Stable isotopes like ^{16}O, ^{18}O, ^{13}C and ^{12}C provide information on past temperatures and other former environmental conditions.

Landnám A localized phase of anthropogenic forest clearance, recognized in pollen diagrams from prehistoric Europe. They usually indicate brief agricultural use followed by abandonment and regeneration of secondary woodland.

Late-glacial (or terminal Pleistocene) The final part of the last glacial stage, dated to between about 16 000 and 11 500 Cal. yr BP.

Little Ice Age   A period of generally colder climate between *c.*AD 1590 and AD 1850, best recorded in northern and central Europe but probably global in extent.

Magnetic susceptibility   An easily measured parameter which gives an indication of the 'magnetizability' of sediment, determined by the abundance of ferro-magnetic minerals. It is particularly useful in studies of lake sediment core correlation.

Mesocratic   The second or early temperate sub-stage of northern European interglacials, when mixed deciduous woodland and brown earth soils became established.

Mesolithic   The Middle Stone Age period of the terminal Pleistocene and early Holocene, whose cultures were reliant on broad spectrum hunting-fishing-gathering economies.

Minerogenic   Clastic sediments which comprise individual mineral particles and are usually classified according to grain size (e.g. sand, clay).

Neolithic   The New Stone Age period characterized by the adoption of agriculture, village life and use of pottery.

Oligocratic   The third or late temperate sub-stage of northern European interglacials, when soil leaching led to podsolization.

Oligotrophic   Lakes, soils or other ecosystems which are nutrient-poor, and which consequently experience low levels of primary productivity by photosynthesizing green plants or algae.

Ombrotrophic   Ecosystems, e.g. raised mires, which are dependent solely on atmospheric sources (mainly precipitation) for their supply of water and nutrients.

Ontogeny   Individual development through time to maturity, e.g. of lake ecosystem.

Palaeoecology   The reconstruction and study of past ecosystems, including the relations between organisms and their environments.

Palaeolimnology   The study of lake history, most importantly from evidence preserved in its bottom sediments.

Palaeolithic   The Old Stone Age period whose Pleistocene cultures ranged from the earliest tool makers at Olduvai in East Africa to the Ice Age hunters of Lascaux.

Plagioclimax   The end-point of a deflected ecosystem succession which is maintained by human interference.

Pleistocene   That period of time, including both glacial and interglacial stages, between *c.*2 million and 11 500 years ago. The late Pleistocene covers the last interglacial–glacial cycle up to the beginning of the Holocene.

Pluvial   A former period of greater wetness, e.g. one of high lake levels; usually, but not always, due to an absolute increase in rainfall.

Protocratic   The initial, pre-temperate sub-stage of northern European interglacials, typically characterized by raw, unleached soils and associated with rapid immigration of pioneer trees such as birch.

Quaternary   The current period of geological time, which comprises the Pleistocene and Holocene, and which began about 2 million years ago. It broadly corresponds to the last Ice Age.

Radiocarbon dating   A **radiometric** method of establishing the age of organic materials up to *c.*40 000 years old, based on the incorporation and decay of the radioactive isotope ^{14}C within organic materials. It is among the most widely applied of all Holocene dating techniques.

Radiometric dating   Those dating techniques based on the principle that radioactive decay of unstable isotopes is time-dependent.

Sclerophyll(ous)   Woody vegetation with hard, evergreen leaves designed to reduce water loss. Often prone to fire.

Sedentism   Living in one main place year round (as opposed to nomadism).

Stadial   An interval of cold climate of shorter duration than a true glacial phase, e.g. Younger Dryas event (13–11 500 Cal. yr BP).

Synchronous   Events that happen simultaneously.

Telocratic   The final, post-temperate sub-stage of northern European interglacials, with declining temperatures and heathland or open conifer woodland.

Tephrochronology   The study of volcanic ash layers (tephra). The deposition of tephra following the eruption of a volcano is effectively a **synchronous** event, and the resulting ash layer provides a marker horizon that can be correlated from one site to another.

Time-transgressive   Processes, such as the spread of a disease, which started earlier in some places and ended later elsewhere.

Uniformitarianism   A basic principle of historical geology and palaeoecology, which, in the sense of 'actualism', implies that past events can be explained in terms of present-day processes and relationships. Uniformitarianism has also – confusingly – been used in a second sense to mean no significant change in the *rate* at which processes take place (or gradualism).

Varves   Annually layered lake sediments.

BIBLIOGRAPHY

Aaby, B. (1976) Cyclic climatic variations in climate over the past 5500 years reflected in raised bogs, *Nature*, 263, 281–4.

Aaby, B. and Berglund, B.E. (1986) Characterization of peat and lake deposits, in B.E. Berglund (ed.), *Handbook of Holocene palaeoecology and palaeohydrology*. Chichester: John Wiley, pp. 231–46.

Aaby, B. and Tauber, H. (1995) Eemian climate and pollen, *Nature*, 376, 27–8.

Abell, P.I., Amegashitsi, L. and Ochumba, P.B.O. (1995) The shells of *Etheria elliptica* as recorders of seasonality at Lake Victoria, *Palaeogeography, Palaeoclimatology, Palaeoecology*, 119, 215–19.

Adams, R.M. (1965) *Land behind Baghdad*. Chicago: University of Chicago Press.

Adams, J.M., Faure, H., Faure-Denard, L., McGlade, J.M. and Woodward, F.I. (1990) Increases in terrestrial carbon storage from the last glacial maximum to the present, *Nature*, 348, 711–14.

Adamson, D.A., Gasse, F., Street, F.A. and Williams, M.A.J. (1980) Late Quaternary history of the Nile, *Nature*, 287, 50–5.

Aitken, M.J. (1990) *Science-based dating in archaeology*. London: Longman.

Aitken, M.J. (1994) Optical dating: a non-specialist review, *Quaternary Science Reviews*, 13, 503–8.

Akazawa, T. (1986) Regional variation in seasonal procurement systems of Jomon hunter-gatherers, in T. Akazawa and C.M. Aikens (eds), *Prehistoric hunter-gatherers in Japan: new research methods*. Tokyo University Museum Bulletin, 27, 73–89.

Allen, B. and Crittenden, R. (1987) Degradation and a pre-capitalist political economy: the case of the New Guinea highlands, in P. Blaikie and H. Brookfield, *Land degradation and society*. London: Methuen, pp. 145–56.

Allen, J.R.M., Huntley, B. and Watts, W.A. (1996) The vegetation and climate of northwest Iberia over the last 14 000 yr, *Journal of Quaternary Science*, 11, 125–47.

Alley, R.B., Meese, D.A., Shuman, C.A., Gow, A.J., Taylor, K.C., Grootes, P.M., White, J.W.C., Ram, M., Waddington, E.D., Mayewski, P.A. and Zielinski, G.A. (1993) Abrupt increase in Greenland snow accumulation at the end of the Younger Dryas event, *Nature*, 362, 527–9.

Allison, T.D., Moeller, R.E. and Davis, M.B. (1986) Pollen in laminated sediments provides evidence for a mid-Holocene forest pathogen outbreak, *Ecology*, 67, 1101–5.

Allott, T.E.H., Harriman, R. and Battarbee, R.W. (1992) Reversibility of acidification at the Round Loch of Glenhead, *Environmental Pollution*, 77, 219–25.

Almendinger, J.E. (1993) A groundwater model to explain past lake levels at Parkers Prairie, Minnesota, USA, *The Holocene*, 3, 105–15.

Almer, B., Dickson, W., Ekstrom, C., Hornstrom, E. and Miller, U. (1974) Effects of acidification on Swedish lakes, *Ambio*, 3, 30–6.

Ammerman, A.J. and Cavalli-Sforza, L.L. (1971) Measuring the rate of spread of early farming in Europe, *Man N.S.*, 6, 674–88.

Amoros, C. and van Urk, G. (1989) Palaeoecological analyses of large rivers: some principles and methods, in G.E. Petts, H. Möller and A.L. Roux (eds), *Historical change of large alluvial rivers*. Chichester: John Wiley, pp. 143–65.

An, Z. (1989) Prehistoric agriculture in China, in D.R. Harris and G.C. Hillman (eds), *Farming and foraging: the evolution of plant exploitation*. London: Unwin Hyman, pp. 643–9.

van Andel, T.H. and Runnels, C.N. (1995) The earliest farmers in Europe, *Antiquity*, 69, 489–500.

van Andel, T.H., Runnels, C.N. and Pope, K.O. (1986) Five thousand years of land use and abuse in the southern Argolid, Greece, *Hesperia,* 55, 103–28.

van Andel, T.H., Jacobsen, T.W., Jolly, J.B. and Lianos, N. (1980) Late Quaternary history of the coastal zone near Franchthi cave, southern Argolid, Greece, *Journal of Field Archaeology*, 7, 389–402.

van Andel, T.H., Zangger, E. and Demitrack, A. (1990) Land use and soil erosion in prehistoric and historical Greece, *Journal of Field Archaeology*, 17, 379–96.

Andersen, P. (1974) *Passages from antiquity to feudalism*. London: New Left Books.

Andersen, S.Th. (1969) Interglacial soil and vegetation development, *Meddeleser fra Dansk Geologisk Forening*, 19, 90–102.

Andersen, S.Th. (1973) The differential pollen productivity of trees and its significance for the interpretation of a pollen diagram from a forested region, in H.J.B. Birks and R.G. West (eds), *Quaternary plant ecology*. Oxford: Blackwell, pp. 109–15.

Anderson, A. (1989) *Prodigious birds. Moas and moa-hunting in prehistoric New Zealand*. Cambridge: Cambridge University Press.

Anderson, N.J. (1989) A whole basin diatom accumulation rate for a small eutrophic lake in Northern Ireland and its palaeoecological implications, *Journal of Ecology*, 77, 926–46.

Anderson, N.J. and Odgaard, B.V. (1994) Recent palaeolimnology of three shallow Danish lakes, *Hydrobiologia*, 275/276, 411–22.

Anderson, N.J., Rippey, B. and Stevenson, A.C. (1993) A comparison of sedimentary and diatom-inferred phosphorus profiles: implications for defining pre-disturbance nutrient conditions, *Hydrobiologia*, 253, 357–66.

Andresen, J.M., Byrd, B.F., Elson, M.D., McGuire, R.H., Mendoza, R.G., Staski, E. and White, J.P. (1981) The deer hunters: Star Carr reconsidered, *World Archaeology*, 13, 31–46.

Andrews, J.T., Davis, P.T., Mode, W.N., Nichols, H. and Short, S.K. (1981) Relative departures in July temperatures in northern Canada for the past 6000 years, *Nature*, 289, 164–7.

Appleby, P.G. and Oldfield, F. (1978) The calculation of lead-210 dates assuming a constant supply of unsupported ^{210}Pb to the sediments, *Catena*, 5, 1–8.

Arnold, C.J. (1984) *Roman Britain and Saxon England. An archaeological study*. London: Croom Helm.

Arnold, D. (1996) *The problem of nature. Environment, culture and European expansion.* Oxford: Blackwell.

Atkinson, T.C., Briffa, K.R. and Coope, G.R. (1987) Seasonal temperatures in Britain during the past 22 000 years reconstructed using beetle remains, *Nature*, 325, 587–92.

Atkinson, T.C., Briffa, K.R., Coope, G.R., Joachim, M.J. and Perzy, D.W. (1986) Climatic calibration of coleopteran data, in B.E. Berglund (ed.), *Handbook of Holocene palaeoecology and palaeohydrology.* Chichester: John Wiley, pp. 851–8.

Baillie, M.G.L. (1995) *A slice through time. Dendrochronology and precision dating.* London: Routledge.

Baillie, M.G.L. and Munro, M.A.R. (1988) Irish tree rings, Santorini and volcanic dust veils, *Nature*, 332, 344–6.

Baker, V.R. (1983) Late-Pleistocene fluvial systems, in S.C. Porter (ed.), *Late-Quaternary environments of the United States. Volume 1, The Late Pleistocene.* London: Longman, pp. 115–29.

Baker, V.R. and Bunker, R.C. (1985) Cataclysmic late Pleistocene flooding from glacial lake Missoula: a review, *Quaternary Science Reviews*, 4, 1–42.

Balaam, N.D., Smith, K. and Wainwright, G.J. (1982) The Shaugh Moor project: fourth report – environment, context and conclusion, *Proceedings of the Prehistoric Society*, 48, 203–78.

Ballais, J.-L. (1995) Alluvial Holocene terraces in eastern Maghreb: climate and anthropogenic controls, in J. Lewin, M.G. Macklin and J.C. Woodward (eds), *Mediterranean Quaternary river environments.* Rotterdam: A.A. Balkema, pp. 183–94.

Bar-Yosef, A. and Belfer-Cohen, A. (1992) From foraging to farming in the Mediterranean Levant, in A.B. Gebauer and T.D. Price (eds), *Transitions to agriculture in prehistory.* Madison, Wisconsin: Prehistory Press, pp. 21–48.

Barber, H.G. and Haworth, E.Y. (1981) *A guide to the morphology of the diatom frustule.* F.B.A. Scientific Publication 44.

Barber, K.E. (1981) *Peat stratigraphy and climate change.* Rotterdam: Balkema.

Barber, K.E. (1993) Peatlands as scientific archives of past biodiversity, *Biodiversity and Conservation*, 2, 474–89.

Barber, K.E., Chambers, F.M., Maddy, D., Stoneman, R. and Brew, J.S. (1994) A sensitive high-resolution record of late Holocene climatic change from a raised bog in northern England, *The Holocene*, 4, 198–205.

Bard, E., Hamelin, B., Arnold, M., Montaggioni, L., Cabioch, G., Faure, G. and Rougerie, F. (1996) Deglacial sea-level record from Tahiti corals and the timing of global meltwater discharge, *Nature*, 382, 241–4.

Bard, E., Hamelin, B., Fairbanks, R.G. and Zindler, A. (1990) Calibration of the ^{14}C timescale over the past 30 000 years using mass spectrometric U-Th ages from Barbados corals, *Nature*, 345, 405–9.

Barker, G. (ed.) (1995a) *The Biferno valley survey. The archaeological and geomorphological record.* London: Leicester University Press.

Barker, G. (1995b) *A Mediterranean valley. Landscape archaeology and* Annales *history in the Biferno valley.* London: Leicester University Press.

Barnett, T.P., Santer, B.D., Jones, P.D., Bradley, R.S. and Briffa, K.R. (1996) Estimates of low frequency natural variability in near-surface air temperature, *The Holocene*, 6, 255–64.

Bartlein, P.J., Prentice, I.C. and Webb, T. (1986) Climate response surfaces from pollen data for some eastern North American taxa, *Journal of Biogeography*, 13, 35–57.

Barton, C.E. and McEllhinny, M.W. (1982) A 10 000 year geomagnetic secular variation record from three Australian maars, *Geophysical Journal of the Royal Astronomical Society*, 67, 465–86.

Baruch, U. and Bottema, S. (1991) Palynological evidence for climatic changes in the Levant ca. 17 000–9000 B.P., in O. Bar-Yosef and F.R. Valla (eds), *The Natufian culture in the Levant*. Ann Arbor, Michigan: International Monographs in Prehistory, pp. 11–20.

Battarbee, R.W. (1978) Observations on the recent history of Lough Neagh and its drainage basin, *Philosophical Transactions of the Royal Society of London, Series B*, 281, 303–45.

Battarbee, R.W. (1984) Diatom analysis and the acidification of lakes, *Philosophical Transactions of the Royal Society of London, Series B*, 305, 451–77.

Battarbee, R.W. (1986) Diatoms in lake sediments, in B.E. Berglund (ed.), *Handbook of Holocene palaeoecology and palaeohydrology*. Chichester: John Wiley, pp. 527–70.

Battarbee, R.W. (1994) Surface water acidification, in N. Roberts (ed.), *The changing global environment*. Oxford: Blackwell, pp. 213–41.

Battarbee, R.W., Flower, R.J., Stevenson, A.C. and Rippey, B. (1985) Lake acidification in Galloway: palaeoecological test of competing hypotheses, *Nature*, 314, 350–2.

Battarbee, R.W., Stevenson, A.C., Rippey, B., Fletcher, C., Natkanski, J., Wik, M. and Flower, R.J. (1989) Causes of lake acidification in Galloway, south-west Scotland: a palaeoecological evaluation of the relative roles of atmospheric contamination and catchment change for two acidified sites with non-afforested catchments, *Journal of Ecology*, 77, 651–72.

Bayliss-Smith, T. (1996) People–plant interactions in the New Guinea highlands: agricultural hearthland or horticultural backwater, in D.R. Harris (ed.), *The origins and spread of agriculture and pastoralism in Eurasia*. London: UCL Press, pp. 499–523.

Becker, B. and Kromer, B. (1993) The continental tree-ring record – absolute chronology, ^{14}C calibration and climatic change at 11 Ka BP, *Palaeogeography, Palaeoclimatology, Palaeoecology*, 103, 67–71.

Beerling, D.J., Birks, H.H. and Woodward, F.I. (1995) Rapid late-glacial atmospheric CO_2 changes reconstructed from the stomatal density record of fossil leaves, *Journal of Quaternary Science*, 10, 379–84.

Behre, K.-E. (1981) The interpretation of anthropogenic indicators in pollen diagrams, *Pollen et Spores*, 23, 225–45.

Behre, K.-E. (ed.) (1986) *Anthropogenic indicators in pollen diagrams*. Rotterdam: Balkema.

Bell, M. (1983) Valley sediments as evidence of prehistoric land-use on the South Downs, *Proceedings of the Prehistoric Society*, 49, 119–50.

Bell, M. and Boardman, J. (eds) (1992) *Past and present soil erosion: archaeological and geographical perspectives.* Oxford: Oxbow Monographs.

Bell, M. and Walker, M.J.C. (1992) *Late Quaternary environmental change. Physical and human perspectives.* London: Longman.

Bellwood, P. (1980) The peopling of the Pacific, *Scientific American*, 243 (5), 74–85.

Bellwood, P. (1987) *The Polynesians. Prehistory of an island people.* London: Thames and Hudson.

Bennett, K.D. (1983a) Postglacial populations expansion of forest trees in Norfolk, U.K., *Nature*, 303, 164–7.

Bennett, K.D. (1983b) Devensian late-glacial and Flandrian vegetational history at Hockham Mere, Norfolk, England. I. Pollen percentages and concentrations, *New Phytologist*, 95, 457–87.

Bennett, K.D. (1989) A provisional map of forest types for the British Isles 5000 years ago, *Journal of Quaternary Science*, 4, 141–4.

Bennett, K.D. (1997) *Evolution and ecology. The pace of life.* Cambridge: Cambridge University Press.

Bennett, K.D., Boreham, S., Sharp, M.J. and Switsur, V.R. (1992) Holocene history of environment, vegetation and human settlement on Catta Ness, Lunnasting, Shetland, *Journal of Ecology*, 80, 241–73.

Bennett, K.D., Simonson, W.D. and Peglar, S.M. (1990) Fire and man in post-glacial woodlands of eastern England, *Journal of Archaeological Science*, 17, 635–42.

Bennett, K.D., Tzedakis, P.C. and Willis, K.J. (1991) Quaternary refugia of north European trees, *Journal of Biogeography* 18, 103–15.

Bennion, H. (1994) A diatom-phosphorus transfer function for shallow, eutrophic ponds in southeast England, *Hydrobiologia*, 275/276, 391–410.

Benson, L., Burdett, J., Lund, S., Kashgarian, M. and Mensing, S. (1997) Nearly synchronous climate change in the northern hemisphere during the last glacial termination, *Nature*, 388, 263–5.

Berger, A.L. (1992) Astronomical theory of paleoclimates and the last glacial–interglacial cycle, *Quaternary Science Reviews*, 11, 571–82.

Berglund, B.E. (ed.) (1986) *Handbook of Holocene palaeoecology and palaeohydrology.* Chichester: John Wiley.

Berglund, B.E. (ed.) (1991) The cultural landscape during 6000 years in southern Sweden, *Ecological Bulletins*, 41.

Berglund, B.E., Birks, H.J.B., Ralska-Jasiewiczowa, M. and Wright, H.E. (eds) (1996) *Palaeoecological events during the last 15 000 years. Regional syntheses of palaeoecological studies of lakes and mires in Europe.* Chichester: John Wiley.

Betancourt, J.L., van Devender, T.R. and Martin, P.S. (eds) (1990) *Packrat middens. The last 40 000 years of biotic change.* Tucson: University of Arizona Press.

Binford, L.R. (1968) Post-Pleistocene adaptations, in S.R. and L.R. Binford (eds), *New perspectives in archaeology.* Chicago: Aldine, pp. 313–41.

Binford, L.R. (1981) *Bones. Ancient men and modern myths.* New York: Academic Press.

Binford, M.W., Brenner, M., Whitmore, T.J., Higuera-Gundy, A., Deevey, E.S. and Leyden, B.W. (1987) Ecosystems, paleoecology and human disturbance in subtropical and tropical America, *Quaternary Science Reviews*, 6, 115–28.

Binford, M.W., Deevey, E.S. and Grisman, T.L. (1983) Palaeolimnology: an historical perspective on lacustrine ecosystems, *Annual Review of Ecology and Systematics*, 14, 255–86.

Bintliff, J. (1981) Archaeology and the Holocene evolution of coastal plains in the Aegean and circum-Mediterranean, in D. Brothwell and G. Dimbleby (eds), *Environmental aspects and coasts and islands*. Oxford: British Archaeological Reports International Series 94, pp. 11–31.

Bircher, W. (1986) Dendrochronology applied in mountain regions, in B.E. Berglund (ed.), *Handbook of Holocene palaeoecology and palaeohydrology*. Chichester: John Wiley, pp. 387–403.

Birks, H.H. (1993) The importance of plant macrofossils in late-glacial climatic reconstructions: an example from western Norway, *Quaternary Science Reviews*, 12, 719–26.

Birks, H.H., Birks, H.J.B., Koland, P.E. and Nioe, D. (eds) (1988) *The cultural landscape: past, present and future*. Cambridge: Cambridge University Press.

Birks, H.J.B. (1981) The use of pollen in the reconstruction of past climates: a review, in T.M.L. Wigley, M.J. Ingram and G. Farmer (eds), *Climate and history*. Cambridge: Cambridge University Press, pp. 111–38.

Birks, H.J.B. (1986a) Late Quaternary biotic changes in terrestrial and lacustrine environments, with particular reference to north-west Europe, in B.E. Berglund (ed.), *Handbook of Holocene palaeoecology and palaeohydrology*. Chichester: John Wiley, pp. 3–65.

Birks, H.J.B. (1986b) Numerical zonation, comparison and correlation of Quaternary pollen-stratigraphical data, in B.E. Berglund (ed.), *Handbook of Holocene palaeoecology and palaeohydrology*. Chichester: John Wiley, pp. 743–74.

Birks, H.J.B. and Birks, H.H. (1980) *Quaternary palaeoecology*. London: Arnold.

Birks, H.J.B. and Mathewes, R.W. (1978) Studies in the vegetational history of Scotland. V. Late Devensian and early Flandrian pollen and macrofossil stratigraphy at Abernethy forest, Inverness-shire, *New Phytologist*, 80, 455–84.

Birks, H.J.B., Deacon, J. and Peglar, S. (1975) Pollen maps for the British Isles 5000 years ago, *Proceedings of the Royal Society of London, Series,* 189, 87–105.

Birks, H.J.B., Line, J.M., Juggins, S., Stevenson, A.C. and ter Braak, C.J.F. (1990) Diatoms and pH reconstruction, *Philosophical Transactions of the Royal Society of London, Series B*, 327, 263–78.

Bishop, P. and Godley, D. (1994) Holocene palaeochannels at SiSatchanalai, north-central Thailand: ages, significance and palaeo-environmental indications, *The Holocene*, 4, 32–41.

Björck, S. and Digerfeldt, G. (1989) Lake Mullsjön – a key site for understanding the final stage of the Baltic Ice Lake east of Mt. Billingen, *Boreas*, 18, 209–19.

Björck, S., Kromer, B., Johnsen, S., Bennike, O., Hammarlund, S., Lemdahl, G., Possnert, G., Rasmussen, T.L., Wohlfarth, B., Hammer, C.U. and Spurk, M. (1996) Synchronized terrestrial-atmospheric deglacial records around the North Atlantic, *Science*, 274, 1156–60.

Black, J.N. (1971) Energy relations in crop production – a preliminary survey, *Annals of Applied Biology*, 67, 272–8.

Black, R.F. (1976) Periglacial features indicative of permafrost: ice and soil wedges, *Quaternary Research*, 6, 3–26.

Blackford, J. (1993) Peat bogs as sources of proxy climatic data: past approaches and future research, in F.M. Chambers (ed.), *Climatic change and human impact on the landscape*. London: Chapman and Hall, pp. 47–56.

Blaikie, P. and Brookfield, H. (1987) *Land degradation and society*. London: Methuen.

Blake, M., Chisholm, B.S., Clark, J.E. and Mudar, K. (1992) Non-agricultural staples and agricultural supplements: early formative subsistence in the Soconusco region, Mexico, in A.B. Gebauer and T.D. Price (eds), *Transitions to agriculture in prehistory*. Madison, Wisconsin: Prehistory Press, pp. 133–51.

Blumler, M.A. (1992) Independent inventionism and recent genetic evidence on plant domestication, *Economic Botany*, 46, 98–111.

Blumler, M.A. and Byrne, R. (1991) The ecological genetics of domestication and the origins of agriculture, *Current Anthropology*, 32, 23–54.

Blunier, T., Chappellaz, J., Schwander, J., Stauffer, B. and Raynaud, D. (1995) Variations in atmospheric methane concentration during the Holocene epoch, *Nature*, 374, 46–9.

Boardman, J., Dearing, J.A. and Foster, I.D.L. (eds) (1990) *Soil erosion on agricultural land*. Chichester: John Wiley.

de Boer, G. (1964) Spurn head: its history and evolution, *Transactions, Institute of British Geographers*, 34, 71–89.

Bökönyi, S. (1976) Development of early stock rearing in the Near East, *Nature*, 264, 19–23.

Bonnefille, R., Roeland, J.C. and Guiot, J. (1990) Temperature and rainfall estimates for the past 40 000 years in equatorial Africa, *Nature*, 346, 347–9.

Bonnett, P.J.P. and Cambray, R.S. (1991) The record of deposition of radionuclides in the sediments of Ponsonby Tarn, Cumbria, *Hydrobiologia*, 214, 63–70.

Bottema, S. (1977) A pollen diagram from the Syrian Anti-Lebanon, *Paléorient*, 3, 259–68.

Bottema, S. (1980) On the history of the walnut (*Juglans regia* L.) in southeastern Europe, *Acta Botanica Neerlandica*, 29, 343–9.

Bottema, S. (1982) Palynological investigations in Greece with special reference to pollen as an indicator of human activity, *Palaeohistoria*, 24, 257–89.

Bottema, S. and Woldring, H. (1984) Late Quaternary vegetation and climate of southwestern Turkey II, *Palaeohistoria*, 26, 123–49.

Bottema, S. and Woldring, H. (1990) Anthropogenic indicators in the pollen record of the eastern Mediterranean, in S. Bottema, G. Entjes-Nieborg and W. van Zeist (eds), *Man's role in the shaping of the eastern Mediterranean landscape*. Rotterdam: A.A. Balkema, pp. 231–64.

Boulton, G.S. (1993) Two cores are better than one, *Nature*, 366, 507–8.

Bowden, M.J., Kates, R.W., Kay, P.A., Riebsame, W.E., Warrick, R.A., Johnson, D.L., Gould, H.E. and Weiner, D. (1981) The effect of climatic fluctuations on human population: two hypotheses, in T.M.L. Wigley, M.J. Ingram and G. Farmer (eds), *Climate and history*. Cambridge: Cambridge University Press, pp. 479–513.

Bowen, H.C. and Fowler, P.J. (1978) *Early land allotment in the British Isles*. Oxford: British Archaeological Reports, British Series 48.

Bowler, J.M. (1981) Australian salt lakes: a palaeohydrologic approach, *Hydrobiologia*, 81/82, 431–44.

Bowler, J.M. and Wasson, R.J. (1984) Glacial age environments of inland Australia, in J.C. Vogel (ed.), *Late Cainozoic palaeoclimates of the southern hemisphere*. Rotterdam: Balkema, pp. 183–208.

Bowler, J.M., Thorne, A.G. and Polach, H.A. (1972) Pleistocene man in Australia: age and significance of the Mungo skeleton, *Nature*, 240, 48–50.

Bradbury, J.P. and Waddington, J.C.B. (1973) The impact of European settlement on Shagawa lake, northeastern Minnesota, U.S.A., in H.J.B. Birks and R.G. West (eds), *Quaternary plant ecology*. Oxford: Blackwell, pp. 289–307.

Bradbury, J.P., Leyden, B., Salgado-Labouriau, M., Lewis, W.M., Schubert, C., Binford, M.W., Frey, D.G., Whitehead, D.R. and Weibazahn, F.H. (1981) Late Quaternary environmental history of Lake Valencia, Venezuela, *Science*, 214, 1299–1305.

Bradley, R. (1984) *Quaternary paleoclimatology*. London: Allen and Unwin.

Bradley, R.S. (ed.) (1991) *Global changes of the past*. Boulder, Colorado: UCAR/Office for Interdisciplinary Earth Studies.

Bradley, R.S. and Jones, P.D. (eds) (1992) *Climate since AD 1500*. London: Routledge.

Bradshaw, R.H.W. (1988) Spatially-precise studies of forest dynamics, in B. Huntley and T. Webb III (eds), *Vegetation history*. Dordrecht: Kluwer, pp. 725–51.

Bradshaw, R.H.W. and Browne, P. (1987) Changing patterns in the post-glacial distribution of *Pinus sylvestris* in Ireland, *Journal of Biogeography*, 14, 237–48.

Bray, W. (1977) From foraging to farming in early Mexico, in J.V.S. Megaw (ed.), *Hunters, gatherers and first farmers beyond Europe*. Leicester: Leicester University Press, pp. 225–50.

Brenner, M. (1983) Paleolimnology of the Peten lake district, Guatemala II. Mayan population density and sediment and nutrient loading of Lake Quexil, *Hydrobiologia*, 103, 205–10.

Briffa, K.R., Bartholin, T.S., Eckstein, D., Jones, P.D., Karlen, W., Schweingruber, F.H. and Zetterberg, P. (1990) A 1 400-year tree-ring record of summer temperatures in Fennoscandia, *Nature*, 346, 434–9.

Broecker, W.S. (1994) Massive iceberg discharges as triggers for global climate change, *Nature*, 372, 421–4.

Broecker, W.S., Bond, G. and Klas, M. (1990) A salt oscillator in the glacial Atlantic? 1. The concept, *Paleoceanography*, 5, 469–77.

Broecker, W.S., Peteet, D. and Rind, D. (1985) Does the ocean-atmosphere have more than one stable mode of operation? *Nature*, 315, 21–5.

Brooks, S.J., Mayle, F.E. and Lowe, J.J. (1997) Chironomid-based lateglacial climatic reconstruction for southeast Scotland, *Journal of Quaternary Science*, 12, 161–7.

Brown, A.G. (1988) The palaeoecology of *Alnus* (alder) and the postglacial history of floodplain vegetation. Pollen percentage and influx data from the West Midlands, United Kingdom, *New Phytologist*, 110, 425–36.

Brown, A.G. (1997) *Alluvial geoarchaeology. Floodplain archaeology and environmental change.* Cambridge: Cambridge University Press.

Brown, A.G. and Barber, K.E. (1985) Late Holocene palaeoecology and sedimentary history of a small lowland catchment in central England, *Quaternary Research*, 24, 87–102.

Brown, A.G. and Keough, M.K. (1994) Holocene floodplain metamorphosis in the Midlands, United Kingdom, *Geomorphology*, 4, 433–5.

Brush, G.S. (1994) The Chesapeake Bay estuarine system, in N. Roberts (ed.), *The changing global environment.* Oxford: Blackwell, pp. 397–416.

Buckland, P. and Dugmore, A. (1991) If this is a refugium, why are my feet so bloody cold? in J.K. Maizels and C. Caseldine (eds), *Environmental change in Iceland: past and present.* Dordrecht: Kluwer, pp. 107–25.

Buckland, P.C. and Sadler, J.P. (1985) The nature of Flandrian alluviation in the Humberhead levels, *East Midlands Geographer*, 8, 239–51.

Buckland, P.C., Ashworth, A.C. and Schwert, D.W. (1995) By-passing Ellis Island: insect immigration to North America, in R.A. Butlin and N. Roberts (eds), *Ecological relations in historical times: human impact and adaptation.* Oxford: Blackwell, pp. 226–44.

Büdel, J. (1982) *Climatic geomorphology.* Princeton: Princeton University Press.

Bulliet, R.W. (1975) *The camel and the wheel.* Cambridge, Massachusetts: Harvard University Press.

Burleigh, R. (1981) W.F. Libby and the development of radiocarbon dating, *Antiquity*, 55, 6–8.

Bush, M. and Colinvaux, P. (1990) A pollen record of a complete glacial cycle from lowland Panama, *Journal of Vegetation Science*, 1, 105–18.

Bush, M.B., Piperno, D.R. and Colinvaux, P.A. (1989) A 6000 year history of Amazonian maize cultivation, *Nature*, 340, 303–5.

Butlin, R.A. and Roberts, N. (eds) (1995) *Ecological relations in historical times: human impact and adaptation.* Oxford: Blackwell.

Butzer, K.W. (1976) *Early hydraulic civilization in Egypt. A study of cultural ecology.* Chicago: University of Chicago Press.

Butzer, K.W. (1982) *Archaeology as human ecology.* Cambridge: Cambridge University Press.

Butzer, K.W. (1984) Long-term Nile flood variation and political discontinuities in Pharaonic Egypt, in J.D. Clark and S.A. Brandt (eds), *From hunters to farmers.* Berkeley: University of California Press, pp. 102–12.

Butzer, K.W., Isaac, G.L., Richardson, J.L. and Washbourn-Kamau, C.K. (1972) Radiocarbon dating of East African lake levels, *Science*, 175, 1069–76.

Campbell, I.D. and McAndrews, J.H. (1993) Forest disequilibrium caused by rapid Little Ice Age cooling, *Nature*, 366, 336–8.

van Campo, E., Duplessy, J.C. and Rossignol-Strick, M. (1982) Climatic conditions deduced from a 150-kyr oxygen isotope-pollen record from the Arabian Sea, *Nature*, 296, 56–9.

van Campo, E., Guiot, J. and Peng, C. (1993) A data-based re-appraisal of the terrestrial carbon budget at the last glacial maximum, *Global and Planetary Change*, 8, 189–202.

Carter, C.E. (1977) The recent history of the chironomid fauna of Lough Neagh, from the analysis of remains in sediment cores, *Freshwater Biology*, 7, 415–23.

Carter, G.F. (1977) A hypothesis suggesting a single origin of agriculture, in C.A. Reed (ed.), *Origins of agriculture*. The Hague: Mouton, pp. 89–133.

Caseldine, C. and Hatton, J. (1993) The development of high moorland on Dartmoor: fire and the influence of Mesolithic activity on vegetation change, in F.M. Chambers (ed.), *Climatic change and human impact on the landscape*. London: Chapman and Hall, pp. 119–31.

Cassels, R. (1984) Faunal extinction and prehistoric man in New Zealand and the Pacific islands, in P.S. Martin and R.G. Klein (eds), *Quaternary extinctions*. Tucson: University of Arizona Press, pp. 741–67.

di Castri, F. and Mooney, H.A. (eds) (1973) *Mediterranean type ecosystems*. Berlin: Springer-Verlag.

Catt, J.A. (1978) The contribution of loess to soils in lowland Britain, in S. Limbrey and J.G. Evans (eds), *The effect of man on the landscape: the lowland zone*. Council for British Archaeology, Research Report 21, pp. 12–20.

Caulfield, S. (1978) Star Carr – an alternative view, *Irish Archaeological Research Forum*, 5, 15–22.

Cavalli-Sforza, L.L. and Cavalli-Sforza, F. (1995) *The great human diasporas: the history of diversity and evolution*. Reading, Massachusetts: Addison-Wesley.

Caviedes, C.N. and Waylen, P.R. (1993) Anomalous westerly winds during ENSO events and colonisation of Easter Island, *Applied Geography*, 13, 123–34.

Chabal, L. (1992) La représentativité paléo-écologique des charbons de bois archéologiques issus du bois de feu, *Bulletin de la Société Botanique de France*, 139, 213–36.

Chambers, F.M. (ed.) (1993) *Climatic change and human impact on the landscape*. London: Chapman and Hall.

Chambers, F.M., Dresser, P.Q. and Smith, A.G. (1979) Radiocarbon dating evidence on the impact of atmospheric pollution on upland peats, *Nature*, 282, 829–31.

Chang, T.-T. (1976) The rice cultures, *Philosophical Transactions of the Royal Society of London, Series B*, 275, 143–55.

Chappell, J. and Polach, H. (1991) Post-glacial sea-level rise from coral record at Huon Peninsula, Papua New Guinea, *Nature*, 349, 147–9.

Chappell, J.M.A. and Grindrod, A. (eds) (1983) *Proceedings of the first CLIMANZ conference*, 2 vols. Canberra: Department of Biogeography and Geomorphology, Australian National University.

Charles, D.F., Whitehead, D.R., Engstrom, D.R., Fry, B.D., Hites, R.A., Norton, S.A., Owen, J.S., Roll, L.A., Schindler, S.C., Smol, J.P., Uutala, A.J., White, J.R. and Wise, R.J. (1987) Paleolimnological evidence for recent acidification of Big Moose Lake, Adirondack Mountains, N.Y. (USA), *Biogeochemistry*, 3, 267–96.

Cherry, J.F. (1981) Pattern and process in the earliest colonisation of the Mediterranean islands, *Proceedings of the Prehistoric Society*, 47, 41–68.

Chester, D.K. and James, P.A. (1991) Holocene alluviation in the Algarve, southern Portugal: the case for an anthropogenic cause, *Journal of Archaeological Science*, 18, 73–88.

Childe, V.G. (1935) *New light on the most ancient East*. London: K. Paul, Trench, Trubner and Co.

Chisholm, B.S. (1989) Variation in diet reconstructions based on stable carbon isotope evidence, in T.D. Price (ed.), *The chemistry of prehistoric human bone*. Cambridge: Cambridge University Press, pp. 10–37.

Chivas, A.R., de Deckker, P. and Shelley, J.M.G. (1986) Magnesium and strontium in non-marine ostracod shells as indicators of palaeosalinity and palaeotemperatures, *Hydrobiologia*, 143, 135–42.

Clapperton, C. (1993) *Quaternary geology and geomorphology of South America*. Amsterdam: Elsevier.

Clark, J.G.D. (1972) Star Carr: a case study in bioarchaeology, *Addison-Wesley Module in Anthropology*, 10, 1–42.

Clark, J.S. (1995) Climate and Indian effects on southern Ontario forests: a reply to Campbell and McAndrews, *The Holocene*, 5, 371–9.

Clark, J.S. and Royall, P.D. (1995) Transformation of a northern hardwood forest by aboriginal (Iroquois) fire: charcoal evidence from Crawford Lake, Ontario, Canada, *The Holocene*, 5, 1–9.

Clark, J.S., Merkt, J. and Müller, H. (1989) Post Glacial fire, vegetation, and human history of the northern Alpine forelands, southwestern Germany, *Journal of Ecology*, 77, 897–925.

Clarke, D. (1976) Mesolithic Europe: the economic basis, in G. de G. Sieveking, I.H. Longworth and K.E. Wilson (eds), *Problems in economic and social archaeology*. London: Duckworth, pp. 449–81.

CLIMAP project members (1976) The surface of ice age earth, *Science*, 191, 1131–7.

CLIMAP project members (1984) The last interglacial ocean, *Quaternary Research*, 21, 123–224.

Cloutman, E.W. (1988) Palaeoenvironments in the Vale of Pickering, *Proceedings of the Prehistoric Society*, 54, 1–36.

Clutton-Brock, J. (1977) Man-made dogs, *Science*, 197, 1340–2.

Clutton-Brock, J. (1987) *A natural history of domesticated mammals*. London: BMNH/Cambridge University Press.

Clymo, R.S. (1991) Peat growth, in L.C.K. Shane and E.J. Cushing (eds), *Quaternary landscapes*. London: Belhaven Press, pp. 76–112.

Coard, M.A., Cousen, S.M., Cuttler, A.H., Dean, H.J., Dearing, J.A., Eglington, T.I., Greaves, A.M., Lacey, K.P., O'Sullivan, P.E., Pickering, D.A., Rhead, M.M., Rodwell, J.K. and Simola, H. (1983) Paleolimnological studies of annually-laminated sediments in Loe Pool, Cornwall, U.K., in J. Merilainen, P. Huttunen and R.W. Battarbee (eds), *Proceedings of the 23rd International Symposium on Paleolimnology, Joensuu, Finland.* The Hague: Dr W. Junk, pp. 185–91.

Cohen, M.N. (1977) *The food crisis in prehistory: overpopulation and the origins of agriculture.* New Haven: Yale University Press.

COHMAP members (1988) Major climatic changes of the last 18 000 years: observations and model simulations, *Science*, 241, 1043–52.

Coles, J.M., Hibbert, F.A. and Orme, B.J. (1973) Prehistoric roads and tracks in Somerset, England. 3. Sweet Track, *Proceedings of the Prehistoric Society,* 39, 256–93.

Colinvaux, P. (1986) *Ecology.* Chichester: John Wiley.

Colinvaux, P. (1987) Amazon diversity in light of the palaeoecological record, *Quaternary Science Reviews,* 6, 93–114.

Connolly, A.P. and Dahl, E. (1970) Maximum summer temperature in relation to the modern and Quaternary distributions of certain arctic-montane species in the British Isles, in D. Walker and R.G. West (eds), *Studies in the vegetation history of the British Isles.* Cambridge: Cambridge University Press, pp. 159–223.

Coope, G.R. (1970) Interpretation of Quaternary insect fossils, *Annual Review of Entomology,* 15, 97–120.

Coope, G.R. (1975) Climatic fluctuations in northwest Europe since the last interglacial, indicated by fossil assemblages of coleoptera, in A.E. Wright and F. Moseley (eds), *Ice ages: ancient and modern.* Liverpool: Seel House Press, pp. 153–68.

Coope, G.R. and Lemdahl, G. (1995) Regional differences in the Lateglacial climate of northern Europe based on coleopteran analysis, *Journal of Quaternary Science,* 10, 391–5.

Coope, G.R. and Lister, A.H. (1987) Late-glacial mammoth skeletons from Condover, Shropshire, England, *Nature,* 330, 472–4.

Cooper, S.R. and Brush, G.S. (1991) Long-term history of Chesapeake Bay anoxia, *Science,* 254, 992–6.

Courty, M.A., Goldberg, P. and MacPhail, R. (1989) *Soils and micromorphology in archaeology.* Cambridge: Cambridge University Press.

Cowgill, U. (1969) The waters of Merom: a study of Lake Huleh, *Archiv für Hydrobiologie,* 66, 249–71.

Crosby, A.W. (1986) *Ecological imperialism. The biological expansion of Europe 900–1900.* Cambridge: Cambridge University Press.

Cumming, B.F. and Smol, J.P. (1993) Development of diatom-based salinity models for paleoclimatic research from lakes in British Columbia (Canada), *Hydrobiologia,* 269/270, 179–96.

Cushing, E.J. (1967) Late-Wisconsin pollen stratigraphy and the glacial sequence in Minnesota, in E.J. Cushing and H.E. Wright (eds), *Quaternary paleoecology.* New Haven: Yale University Press, pp. 59–88.

Darch, J.P. (1988) Drained field agriculture in tropical Latin America: parallels from past to present, *Journal of Biogeography,* 15, 87–96.

Davidson, D.A. (1980) Erosion in Greece during the first and second millennia BC, in R.A. Cullingford, D.A. Davidson and J. Lewin (eds), *Timescales in geomorphology*. Chichester: John Wiley, pp. 143–58.

Davidson, D.A. and Shackley, M.L. (eds) (1976) *Geoarchaeology*. London: Duckworth.

Davidson, J.M. (1984) *The prehistory of New Zealand*. Auckland: Longman Paul.

Davis, M.B. (1976a) Pleistocene biogeography of temperate deciduous forests, *Geoscience and Man*, 13, 13–26.

Davis, M.B. (1976b) Erosion rates and land use history in southern Michigan, *Environmental Conservation*, 3, 139–48.

Davis, R.B. (1987) Paleolimnological diatom studies of acidification lakes by acid rain: an application of Quaternary Science, *Quaternary Science Reviews*, 6, 147–64.

de Deckker, P. (1981) Ostracods of athalassic saline lakes. A review, *Hydrobiologia*, 81, 131–44.

Dearing, J. (1986) Core correlation and total sediment influx, in B.E. Berglund (ed.), *Handbook of Holocene palaeoecology and palaeohydrology*. Chichester: John Wiley, pp. 247–72.

Dearing, J. (1994) Reconstructing the history of soil erosion, in N. Roberts (ed.), *The changing global environment*. Oxford: Blackwell, pp. 242–61.

Dearing, J.A. (1983) Changing patterns of sediment accumulation in a small lake in Scania, southern Sweden, in J. Merilainen, P. Huttunen and R.W. Battarbee (eds), *Proceedings of the 23rd International Symposium on Paleolimnology, Joensuu, Finland*. The Hague: Dr W. Junk, pp. 59–64.

Dearing, J.A., Häkansson, H., Liedberg-Jönsson, B., Persson, A., Skansjö, S., Widholm, D. and El-Daoushy, F. (1987) Lake sediments used to quantify the erosional response to land use change in southern Sweden, *Oikos*, 50, 60–78.

Deevey, E.S. (1942) Studies on Connecticut lake sediments III. The biostratonomy of Linsley Pond, *American Journal of Science*, 240, 235–64, 313–24.

Deevey, E.S. (1969) Coaxing history to conduct experiments, *BioScience*, 19, 40–3.

Deevey, E.S. (1984) Stress, strain, and stability of lacustrine ecosystems, in E. Haworth and J.W.G. Lund (eds), *Lake sediments and environmental history*. Leicester: Leicester University Press, pp. 208–29.

Deevey, E.S., Gross, M.S., Hutchinson, G.E. and Kraybill, H.L. (1954) The natural ^{14}C contents of materials from hard-water lakes, *Proceedings of the National Academy of Sciences*, 40, 285–8.

Deevey, E.S., Rice, D.S., Rice, P.M., Vaughan, H.H., Brenner, M. and Flannery, M.S. (1979) Mayan urbanism: impact on a tropical karst landscape, *Science*, 206, 298–306.

Delcourt, P.A. and Delcourt, H.R. (1987) *Long term forest dynamics in the temperate zone*. Berlin/New York: Springer-Verlag.

Denbow, J.R. and Wilmsen, E.N. (1986) Advent and course of pastoralism in the Kalahari, *Science*, 234, 1509–15.

Dennell, R.W. (1983) The hunter-gatherer/agricultural frontier in prehistoric temperate Europe, in S.W. Green and S.M. Perlman (eds),

The archaeology of frontiers and boundaries. New York: Academic Press, pp. 113–39.

Denton, G.H. and Hughes, T.J. (eds) (1981) *The last great ice sheets*. New York: John Wiley.

Diamond, J. (1997) *Guns, germs and steel. A short history of everybody for the last 13 000 years*. London: Jonathan Cape.

Diamond, J.M. (1994) The last people alive, *Nature*, 370, 331–2.

Dimbleby, G.W. (1978) Changes in ecosystems through forest clearance, in J.G. Hawkes (ed.), *Conservation and agriculture*. London: Duckworth, pp. 3–16.

Dimbleby, G.W. (1984) Anthropogenic changes from neolithic through medieval times, *New Phytologist*, 98, 57–72.

Dixit, S.S., Dixit, A.S. and Smol, J.P. (1989) Lake acidification recovery can be monitored using chrysophycean microfossils, *Canadian Journal of Fisheries and Aquatic Sciences*, 46, 1309–12.

Dixit, S.S., Dixit, A.S., Smol, J.P. and Keller, W. (1995) Reading the records stored in the lake sediments: a method of examining the history and extent of industrial damage to lakes, in J.M. Gunn (ed.), *Restoration and recovery of an industrial region*. New York: Springer-Verlag, pp. 33–44.

Dixit, S.S., Smol, J.P., Kingston, J.C. and Charles, D.F. (1992) Diatoms: powerful indicators of environmental change, *Environmental Science and Technology*, 26, 23–33.

Dodgshon, R.A. (1988) The ecological basis of Highland peasant farming 1500–1800 AD, in H.H. Birks, H.J.B. Birks, P.E. Koland and D. Nioe (eds), *The cultural landscape: past, present and future*. Cambridge: Cambridge University Press, pp. 139–51.

Dolukhanov, P.M. and Khotinsky, N.A. (1984) Human cultures and the natural environment in the USSR during the Mesolithic and Neolithic, in A.A. Velichko (ed.), *Late Quaternary environments in the Soviet Union*. London: Longman, pp. 319–27.

Donkin, R.A. (1997) A 'servant of two masters?', *Journal of Historical Geography*, 23, 247–66.

Douglas, I. (1967) Man, vegetation and the sediment yields of rivers, *Nature*, 215, 25–8.

Douglas, I. and Spencer, T. (eds) (1985) *Environmental change and tropical geomorphology*. London: George Allen and Unwin.

Drew, D.P. (1982) Environmental archaeology and karstic terrains: the example of the Burren, Co. Clare, Ireland, in M. Bell and S. Limbrey (eds), *Archaeological aspects of woodland ecology*. Oxford: British Archaeological Reports, International Series 146, pp. 115–27.

Drewry, D. (1985) *Glacial geologic processes*. London: Arnold.

Dugmore, A. and Buckland, P. (1991) Tephrochronology and Late Holocene soil erosion in South Iceland, in J.K. Maizels and C. Caseldine (eds), *Environmental change in Iceland: past and present*. Dordrecht: Kluwer, pp. 147–59.

Duller, G.A.T. (1996) Recent developments in luminescence dating of Quaternary sediments, *Progress in Physical Geography*, 20, 127–45.

Dumayne, L. and Barber, K.E. (1994) Pollen diagrams from Hadrian's Wall, *The Holocene*, 4, 165–73.

Dumond, D.E. (1980) The colonization of the Arctic, in A. Sherratt (ed.), *The Cambridge Encyclopaedia of Archaeology*. Cambridge: Cambridge University Press, pp. 361–4.

Dury, G.H. (1964) Principles of underfit streams, *U.S. Geological Survey Professional Paper* 452–A.

Dury, G.H. (1965) Theoretical implications of underfit streams, *U.S. Geological Survey Professional Paper* 452–C.

Dyke, A.S. and Prest, V.K. (1987) Late Wisconsinan and Holocene history of the Laurentide ice sheet, *Géographie Physique et Quaternaire*, 41, 237–64.

Edmondson, W.T. (1974) The sedimentary record of the eutrophication of Lake Washington, *Proceedings of the National Academy of Science, USA*, 71, 5093–5.

Edwards, K.J. (1979) Palynological and temporal inference in the context of prehistory, with special reference to the evidence from lake and peat deposits, *Journal of Archaeological Science*, 6, 255–70.

Edwards, K.J. (1982) Man, space and the woodland edge – speculations on the detection and interpretation of human impact in pollen profiles, in M. Bell and S. Limbrey (eds), *Archaeological aspects of woodland ecology*. Oxford: British Archaeological Reports, International Series 146, pp. 5–22.

Edwards, K.J. (1993) Models of mid-Holocene forest farming for northwest Europe, in F.M. Chambers (ed.), *Climatic change and human impact on the landscape*. London: Chapman and Hall, pp. 133–45.

Edwards, K.J. and MacDonald, G.M. (1991a) Holocene palynology: I Principles, population and community ecology, palaeoclimatology, *Progress in Physical Geography*, 15, 261–89.

Edwards, K.J. and MacDonald, G.M. (1991b) Holocene palynology: II Human influence and vegetation change, *Progress in Physical Geography*, 15, 364–91.

Edwards, K.J. and Rowntree, K.M. (1980) Radiocarbon and palaeoenvironmental evidence for changing rates of erosion at a Flandrian stage site in Scotland, in R.A. Cullingford, D.A. Davidson and J. Lewin (eds), *Timescales in geomorphology*. Chichester: John Wiley, pp. 207–23.

Edwards, K.J. and Whittington, G. (1992) Palynological evidence for the growing of *Cannabis sativa* L. (hemp), and variations on *Cannabis* pollen representation, *Transactions, Institute of British Geographers N.S.*, 15, 60–9.

Edwards, K.J., Buckland, P.C., Blackford, J.J., Dugmore, A.J. and Sadler, J.P. (1994) The impact of tephra: proximal and distal studies of Icelandic eruptions, *Münchener Geographische Abhandlungen*, B12, 79–99.

Eisma, D. (1978) Stream deposition and erosion by the eastern shore of the Aegean, in W.C. Brice (ed.), *The environmental history of the Near and Middle East since the last Ice Age*. London: Academic Press, pp. 67–81.

Elias, S.A., Short, S.K., Nelson, C.H. and Birks, H.H. (1996) Life and times of the Bering land bridge, *Nature*, 382, 60–3.

Elwell, H.A. and Stocking, M.A. (1984) Estimating soil life-span for conservation planning, *Tropical Agriculture*, 61, 148–50.

Emiliani, C. (1980) Ice sheets and ice melts, *Natural History*, 89, 82–91.

Engstrom, D.R. and Nelson, S.R. (1990) Palaeosalinity from trace metals in fossil ostracods compared with observational records at Devils Lake, North Dakota, U.S.A., *Palaeogeography, Palaeoclimatology, Palaeoecology*, 83, 295–312.

Engstrom, D.R. and Swain, E.B. (1997) Recent declines in atmospheric mercury deposition in the Upper Midwest, *Environmental Science and Technology*, 31, 960–7.

Engstrom, D.R. and Wright, H.E. (1984) Chemical stratigraphy of lake sediments as a record of environmental change, in E.Y. Haworth and J.W.G. Lund (eds), *Lake sediments and environmental history*. Leicester: Leicester University Press, pp. 11–67.

Eronen, M. (1983) Late Weichselian and Holocene shore displacement in Finland, in D.E. Smith and A.G. Dawson (eds), *Shorelines and isostasy*. London: Academic Press, pp. 183–208.

Eugster, H.P. and Kelts, K. (1983) Lacustrine chemical sediments, in A.S. Goudie and K. Pye (eds), *Chemical sediments and geomorphology*. London: Academic Press, pp. 321–68.

Evans, J.G. (1972) *Land snails in archaeology*. London: Seminar Press.

Evans, J.G. (1993) The influence of human communities on the English chalklands from the Mesolithic to the Iron Age: the molluscan evidence, in F.M. Chambers (ed.), *Climatic change and human impact on the landscape*. London: Chapman and Hall, pp. 147–56.

Faegri, K. and Iverson, J. (1989) *Textbook of pollen analysis*, 4th edn. Chichester: John Wiley.

Fairbanks, R.G. (1989) A 17 000-year glacio-eustatic sea level record: influence of glacial melting rates on the Younger Dryas 'event' and deep-ocean circulation, *Nature*, 342, 637–42.

Fairbridge, R.W. (1982) The Pleistocene–Holocene boundary, *Quaternary Science Reviews*, 1, 215–44.

Falkowski, P.G. (1997) Evolution of the nitrogen cycle and its influence on the biological sequestration of CO_2 in the ocean, *Nature*, 387, 272–5.

Fall, P.L. (1990) Deforestation in southern Jordan: evidence from fossil hyrax middens, in S. Bottema, G. Entjes-Nieborg and W. van Zeist (eds), *Man's role in the shaping of the eastern Mediterranean landscape*. Rotterdam: A.A. Balkema, pp. 271–81.

Farnsworth, P., Brady, J.E., DeNiro, M.J. and MacNeish, R.S. (1985) A re-evaluation of the isotopic and archaeological reconstructions of diet in the Tehuacán valley, *American Antiquity*, 50, 102–16.

Faure, H., Manguin, E. and Nydal, R. (1963) Formations lacustres du Quaternaire supérieur du Niger oriental: diatomites et âges absolus, *Bulletin du Bureau de Recherches de Géologie et Minières*, 3, 41–63.

Ferguson, C.W. (1968) Bristlecone pine: science and esthetics, *Science*, 159, 839–46.

Field, M.H., Huntley, B. and Müller, H. (1994) Eemian climate fluctuations observed in a European pollen record, *Nature*, 371, 779–83.

Figge, R.A. and White, J.W.C. (1995) High-resolution Holocene and late glacial atmospheric CO_2 record: variability tied to changes in thermohaline circulation, *Global Biogeochemical Cycle*, 9, 391–403.

Finlayson, W.D. and Byrne, R. (1975) Investigations of Iroquoian settlement at Crawford Lake, Ontario – a preliminary report, *Ontario Archaeology*, 25, 31–6.

Fleming, A. (1978) The prehistoric landscape of Dartmoor. Part 1: South Dartmoor, *Proceedings of the Prehistoric Society*, 44, 97–123.

Fleming, A. (1983) The prehistoric landscape of Dartmoor. Part 2: North and East Dartmoor, *Proceedings of the Prehistoric Society*, 49, 195–242.

Flenley, J.R. (1979a) *The equatorial rainforest: a geological history.* London: Butterworth.

Flenley, J.R. (1979b) The Late Quaternary vegetational history of the equatorial mountains, *Progress in Physical Geography*, 3, 488–509.

Flenley, J.R. (1988) Palynological evidence for land use changes in South-East Asia, *Journal of Biogeography*, 15, 87–96.

Flenley, J.R. and King, S.M. (1984) Late Quaternary pollen records from Easter Island, *Nature*, 307, 47–50.

Flenley, J.R., King, S.A., Jackson, J., Chew, C., Teller, J.T. and Prentice, M.E. (1991) The Late Quaternary vegetational and climatic history of Easter Island, *Journal of Quaternary Science*, 6, 85–115.

Flower, R.J. and Battarbee, R.W. (1983) Diatom evidence for recent acidification of two Scottish lochs, *Nature*, 305, 130–3.

Flower, R.J., Mackay, A.W., Rose, N.L., Boyle, J.L., Dearing, J.A., Appleby, P.G., Kuzmina, A.E. and Granina, L.Z. (1995) Sedimentary records of recent environmental change in Lake Baikal, Siberia, *The Holocene*, 5, 323–7.

Flower, R.J., Stevenson, A.C., Dearing, J.A., Foster, I.D.L., Airey, A., Rippey, B., Wilson, J.P.F. and Appleby, P.G. (1989) Catchment disturbance inferred from palaeolimnological studies of three contrasted sub-humid environments in Morocco, *Journal of Paleolimnology*, 1, 293–322.

Foley, J.A., Kutzbach, J.E., Coe, M.T. and Levis, S. (1994) Feedbacks between climate and boreal forests during the Holocene epoch, *Nature*, 371, 52–4.

Folk, R.L. (1974) *Petrology of sedimentary rocks.* Austin, Texas: Hemphill.

Fontes, J.-Ch. and Gasse, F. (1991) PALYDAF (Palaeohydrology in Africa) program: objectives, methods, major results, *Palaeogeography, Palaeoclimatology, Palaeoecology*, 84, 191–215.

Fontes, J.-Ch., Gasse, F., Camara, E., Millet, B., Saliège, J.F. and Steinberg, M. (1985) Late Holocene changes in Lake Abhé hydrology (Ethiopia-Djibouti), *Zeitschrift für Gletscherkunde*, 21, 89–96.

Fontugne, M., Arnold, M., Labeyrie, L., Paterne, M., Calvert, S.E. and Duplessy, J.-C. (1994) Paleoenvironment, sapropel chronology and Nile River discharge during the last 20 000 years as indicated by deep-sea sediment records in the eastern Mediterranean, in O. Bar-Yosef and R.S. Kra (eds), *Late Quaternary chronology and paleoclimates of the eastern Mediterranean*. Tucson, Arizona: Radiocarbon, pp. 75–88.

Foster, I.D.L., Dearing, J.A. and Grew, R. (1988) Lake catchments: an evaluation of their contribution to studies of sediment yield and delivery processes, *Hydrological Sciences Publication*, 174, 413–24.

Foster, I.D.L., Dearing, J.A., Simpson, A., Carter, A.D. and Appleby, P.G. (1985) Lake catchment based studies of erosion and denudation in the Merevale catchment, Warwickshire, UK, *Earth Surface Processes and Landforms*, 10, 45–68.

Fowler, P.J. (1983) *The farming of prehistoric Britain*. Cambridge: Cambridge University Press.

Frey, D.G. (1986) Cladocera analysis, in B.E. Berglund (ed.), *Handbook of Holocene palaeoecology and palaeohydrology*. Chichester: John Wiley, pp. 667–92.

Fritz, H.C. (1991) *Reconstructing large-scale climatic patterns from tree-ring data*. Tucson: University of Arizona Press.

Fritz, S.C., Engstrom, D.R. and Haskell, B.J. (1994) 'Little Ice Age' aridity in the North American Great Plains: a high-resolution reconstruction of salinity fluctuations from Devils Lake, North Dakota, USA, *The Holocene*, 4, 69–73.

Fritz, S.C., Juggins, S., Battarbee, R.W. and Engstrom, D.R. (1991) Reconstruction of past changes in salinity and climate using a diatom based transfer function, *Nature*, 352, 706–8.

Fujii, S. and Fuji, N. (1967) Postglacial sea level in the Japanese islands, *Journal of Geosciences, Osaka City University*, 10, 43–51.

Fullagar, R.L.K., Prince, D.M. and Head, L.M. (1996) Early human occupation of northern Australia: archaeology and thermoluminescence dating of Jinmium rock-shelter, Northern Territory, *Antiquity*, 70, 751–73.

Furley, P.A., Munro, D.M., Darch, J.P. and Randall, R.R. (1995) Human impact on the Wetlands of Belize, Central America, in R.A. Butlin and N. Roberts (eds), *Ecological relations in historical times: human impact and adaptation*. Oxford: Blackwell, pp. 280–307.

Gasse, F. (1987) Diatoms for reconstructing palaeoenvironments and palaeohydrology in tropical semi-arid zones, *Hydrobiologia*, 154, 127–63.

Gasse, F. and van Campo, E. (1994) Abrupt post-glacial climatic events in West Asia and North Africa monsoon domains, *Earth and Planetary Science Letters*, 126, 435–56.

Gasse, F., Téhet, R., Durand, A., Gibert, E. and Fontes, J.-Ch. (1990) The arid–humid transition in the Sahara and the Sahel during the last deglaciation, *Nature*, 346, 141–6.

Gautier, A. (1980) Contributions to the archaeozoology of Egypt, in F. Wendorf and R. Schild (eds), *Prehistory of the eastern Sahara*. New York: Academic Press, pp. 317–44.

Gaven, C., Hillaire-Marcel, C. and Petit-Maire, N. (1981) A Pleistocene lacustrine episode in southeastern Libya, *Nature*, 290, 131–3.

Gear, A.J. and Huntley, B. (1991) Rapid changes in the range limits of Scots Pine 4000 years ago, *Science*, 251, 544–7.

Geist, V. (1989) Did large predators keep humans out of North America?, in J. Clutton-Brock (ed.), *The walking larder. Patterns of domestication, pastoralism and predation*. London: Unwin Hyman, pp. 282–94.

Gerrard, J. (1985) Soil erosion and landscape stability in southern Iceland: a tephrochronological approach, in K.S. Richards, R.R. Arnett and S. Ellis (eds), *Geomorphology and soils*. London: George Allen and Unwin, pp. 78–95.

Gerrard, J. (1991) An assessment of some of the factors involved in recent landscape change in Iceland, in J.K. Maizels and C. Caseldine (eds), *Environmental change in Iceland: past and present*. Dordrecht: Kluwer, pp. 237–53.

Gillespie, R. (1984) *Radiocarbon user's handbook*. Oxford: Oxford University Committee for Archaeology Monograph no. 3.

Gillespie, R., Horton, D.R., Ladd, P., Macumber, P.G., Rich, T.H., Thorne, R. and Wright, R.V.S. (1978) Lancefield Swamp and the extinction of the Australian megafauna, *Science*, 200, 1044–8.

Gillespie, R., Street-Perrott, F.A. and Switsur, R. (1983) Post-glacial arid episodes in Ethiopia have implications for climate prediction, *Nature,* 306, 681–3.

Gillieson, D., Gorecki, P., Head, J. and Hope, G. (1986) Soil erosion and agricultural history in the central highlands of Papua New Guinea, in V. Gardiner (ed.), *Proceedings of 1st International Conference on Geomorphology*, Manchester, UK, Part II. Chichester: John Wiley, pp. 507–22.

Gimingham, C.H. and de Smidt, J.T. (1983) Heaths as natural and semi-natural vegetation, in W. Holzner, M.J.A. Werger and I. Ikusimar (eds), *Man's impact on vegetation*. The Hague: Dr W. Junk, pp. 185–99.

Girling, M.A. and Greig, J. (1985) A first fossil record for *Scolytus scolytus* (F.) (elm bark beetle): its occurrence in elm decline deposits from London and their implications for Neolithic elm decline, *Journal of Archaeological Science*, 12, 347–51.

Glacken, C.J. (1967) *Traces on the Rhodian shore*. Berkeley: University of California Press.

Glover, I.C. (1980) Agricultural origins in East Asia, in A. Sherratt (ed.), *The Cambridge Encyclopaedia of Archaeology*. Cambridge: Cambridge University Press, pp. 152–61.

Glover, I.C. and Higham, C.F.W. (1996) New evidence for early rice cultivation in South, Southeast and East Asia, in D.R. Harris (ed.), *The origins and spread of agriculture and pastoralism in Eurasia*. London: UCL Press, pp. 413–41.

Golson, J. (1977) No room at the top: agricultural intensification in the New Guinea highlands, in J. Allen, J. Golson and R. Jones (eds), *Sunda and Sahul*. London: Academic Press, pp. 601–38.

Golson, J. and Hughes, P.J. (1980) The appearance of plant and animal domestication in New Guinea, *Journal de la Société des Océanistes*, 36, 294–303.

Goodfriend, G.A. (1992) The use of land snail shells in paleo-environmental reconstruction, *Quaternary Science Reviews*, 11, 665–85.

Goodfriend, G.A., Cameron, R.A.D. and Cook, L.M. (1994) Fossil evidence of recent human impact on the land snail fauna of Madeira, *Journal of Biogeography*, 21, 309–20.

Godwin, H. (1975) *The history of the British flora*, 2nd edn. Cambridge: Cambridge University Press.

Godwin, H. (1978) *Fenland: its ancient past and uncertain future*. Cambridge: Cambridge University Press.

Göransson, H. (1986) Man and the forests of nemoral broad-leaved trees during the Stone Age, *Striae*, 24, 143–52.

Goslar, T., Arnold, M., Bard, E., Kuc, T., Pazdur, M.F., Ralska-Jasiewiczowa, M., Rózanski, K., Tisnerat, N., Walanus, A., Wicik, B. and Wieckowski, K. (1995) High concentration of atmospheric ^{14}C during the Younger Dryas cold episode, *Nature*, 377, 414–17.

Goudie, A. (1993) *The human impact*, 4th edn. Oxford: Blackwell.

Goudie, A. (ed.) (1994) *Geomorphological techniques*, 2nd edn. London: Routledge.

Goudie, A.S. (1983) The arid earth, in R. Gardner and H. Scoging (eds), *Megageomorphology*. Oxford: Clarendon Press, pp. 152–71.

Grattan, J. and Brayshay, M. (1995) An amazing and portentous summer: environmental and social responses in Britain to the 1783 eruption of an Iceland volcano, *Geographical Journal*, 161, 125–34.

Grayson, D.K. (1977) Pleistocene avifaunas and the overkill hypothesis, *Science*, 195, 691–2.

Gregory, K.J., Lewin, J. and Thornes, J.B. (eds) (1987) *Palaeohydrology in practice. A river basin analysis*. Chichester: John Wiley.

Greig, J. (1982) Past and present lime woods of Europe, in M. Bell and S. Limbrey (eds), *Archaeological aspects of woodland ecology*. Oxford: British Archaeological Reports, International Series 146, pp. 23–55.

Grosswald, M.G. (1980) Late Weichselian ice sheet of northern Eurasia, *Quaternary Research*, 13, 1–32.

Grove, A.T. and Warren, A. (1968) Quaternary landforms and climate on the south side of the Sahara, *Geographical Journal*, 134, 194–208.

Grove, J.M. (1979) The glacial history of the Holocene, *Progress in Physical Geography*, 3, 1–54.

Grove, J.M. (1988) *The Little Ice Age*. London: Routledge.

Grove, J.M. and Battagel, A. (1981) Tax records as an index of Little Ice Age environmental and economic deterioration from Sunnfjord Fogderi, Norway, in C. Delano Smith and M. Parry (eds), *Consequences of climatic change*. Nottingham: Nottingham University Press, pp. 70–87.

Grove, R.H. (1995) *Green imperialism: colonial expansion, tropical island Edens and the origins of environmentalism, 1600–1860*. Cambridge: Cambridge University Press.

Guilderson, T.P., Fairbanks, R.G. and Rubenstone, J.L. (1994) Tropical temperature variations since 20 000 years ago: modulating interhemispheric climate change, *Science*, 263, 663–5.

Gulliksen, S., Birks, H.H., Possnert, G. and Mangerud, J. (1998) The calendar age of the Younger Dryas – Holocene transition at Kråkenes, western Norway, *The Holocene*, *

Hagelberg, E., Quevado, S., Turbon, D. and Clegg, J.B. (1994) DNA from ancient Easter Islanders, *Nature*, 369, 25–6.

Hamilton, A.C. (1982) *Environmental history of East Africa*. London: Academic Press.

van der Hammen, T. (1974) The Pleistocene changes of vegetation and climate in tropical South America, *Journal of Biogeography*, 1, 3–26..

Hannon, G.E. and Bradshaw, R.H.W. (1989) Recent vegetation dynamics on two Connemara lake islands, western Ireland, *Journal of Biogeography*, 16, 75–81.

Hansen, J.M. (1991) *The palaeoethnobotany of Franchthi cave. Excavations at Franchthi cave, Greece, Fascicle 7*. Bloomington/Indianapolis: Indiana University Press.

Harlan, J.R. (1967) A wild wheat harvest in Turkey, *Archaeology*, 20, 197–201.

Harlan, J.R. (1971) Agricultural origins: centres and noncentres, *Science*, 174, 468–74.

Harlan, J.R. (1975) *Crops and man*. Madison, Wisconsin: American Society of Agronomy.

Harlan, J.R. (1995) *The living fields. Our agricultural heritage*. Cambridge: Cambridge University Press.

Harlan, J.R. and Stemler, A. (1976) The races of sorghum in Africa, in J.R. Harlan, J.M.J. de Wet and A.B.L. Stemler (eds), *Origins of African plant domestication*. The Hague: Mouton, pp. 465–78.

Harris, D.R. (1984) Ethnohistorical evidence for the exploitation of wild grasses and forbs: its scope and archaeological implications, in W. van Zeist and W.A. Casparie (eds), *Plants and ancient man*. Rotterdam: A.A. Balkema, pp. 63–9.

Harris, D.R. (1990) Vavilov's concept of centres of origin of cultivated plants: its genesis and its influence of the study of agricultural origins, *Biological Journal of the Linnean Society*, 39, 7–16.

Harris, D.R. (ed.) (1996a) *The origins and spread of agriculture and pastoralism in Eurasia*. London: UCL Press.

Harris, D.R. (1996b) Domesticatory relationships between people, plants and animals, in R. Ellen and K. Fukui (eds), *Redefining nature: ecology, culture and domestication*. Oxford: Berg, pp. 437–63.

Harris, D.R. and Hillman, G.C. (eds) (1989) *Farming and foraging: the evolution of plant exploitation*. London: Unwin Hyman.

Harris, S.A. (1986) *The permafrost environment*. London: Croom Helm.

Haskell, B.J., Engstrom, D.R. and Fritz, S.C. (1996) Late Quaternary paleohydrology in the North American Great Plains inferred from the geochemistry of endogenic carbonate and fossil ostracods from Devils Lake, North Dakota, USA, *Palaeogeography, Palaeoclimatology, Palaeoecology*, 124, 179–93.

Hassan, F. (1986) Desert environment and origins of agriculture in Egypt, *Norwegian Archaeological Review*, 19, 63–76.

Hassan, F. (1988) Holocene Nile floods and their implications for the origins of Egyptian agriculture, in J. Bower and D. Lubell (eds), *Prehistoric cultures and environments in the Late Quaternary of Africa*. Oxford: British Archaeological Reports, International Series 405, pp. 1–17.

Hassan, F. (1993) Population ecology and civilization in ancient Egypt, in C.L. Crumley (ed.), *Historical ecology. Cultural knowledge and changing landscapes*. Santa Fe: School of American Research Press, pp. 155–81.

Hather, J.G. (ed.) (1994) *Tropical archaeobotany: applications and new developments*. London: Routledge.

Hawkes, J.G. (1989) The domestication of roots and tubers in the American tropics, in D.R. Harris and G.C. Hillman (eds), *Farming and foraging: the evolution of plant exploitation*. London: Unwin Hyman, pp. 481–503.

Haworth, E.Y. (1977) The sediments of Lake George (Uganda) V. The diatoms in relation to ecological history, *Archiv für Hydrobiologie*, 80, 200–15.

Haynes, C.V. (1991) Geoarchaeological and paleohydrological evidence for a Clovis-age drought in North America and its bearing on extinction, *Quaternary Research*, 35, 438–50.

Hays, J.D., Imbrie, J. and Shackleton, N.J. (1976) Variations in the earth's orbit: pacemaker of the ice ages, *Science*, 235, 1156–67.

Heaton, T.H.E., Holmes, J.A. and Bridgwater, N.D. (1995) Carbon and oxygen isotope variations among lacustrine ostracods: implications for palaeoclimatic studies, *The Holocene*, 5, 428–34.

Hedges, R.E.M. (1991) AMS dating; present status and potential applications, in J.J. Lowe (ed.), *Radiocarbon dating: recent applications and future potential.* (*Quaternary Proceedings*, 1.) London: Quaternary Research Association, pp. 5–10.

Helbaek, H. (1969) Palaeo-ethnobotany, in D. Brothwell and E. Higgs (eds), *Science and archaeology*, 2nd edn. London: Thames and Hudson, pp. 206–14.

Heusser, C.J. (1993) Late-glacial of southern South America, *Quaternary Science Reviews*, 12, 345–50.

Heusser, C.J. and Streeter, S.S. (1980) A temperature and precipitation record of the past 16 000 years in southern Chile, *Science*, 210, 1345–7.

Higgs, E.S. (ed.) (1972) *Papers in economic prehistory.* Cambridge: Cambridge University Press.

Higgs, E.S. (ed.) (1975) *Palaeoeconomy.* Cambridge: Cambridge University Press.

Higgs, E.S. and Vita-Finzi, C. (1972) Prehistoric economies: a territorial approach, in E.S. Higgs (ed.), *Papers in economic prehistory.* Cambridge: Cambridge University Press, pp. 27–36.

Hillman, G. (1984) Interpretation of archaeological plant remains: the application of ethnographic models from Turkey, in W. van Zeist and W.A. Casparie (eds), *Plants and ancient man.* Rotterdam: A.A. Balkema, pp. 1–42.

Hillman, G. (1996) Late Pleistocene changes in wild plant-foods available to hunter-gatherers of the northern Fertile Crescent: possible preludes to cereal cultivation, in D.R. Harris (ed.), *The origins and spread of agriculture and pastoralism in Eurasia.* London: UCL Press, pp. 159–203.

Hillman, G. and Davies, M.S. (1990) Measured domestication rates in wild wheats and barley under primitive cultivation, and their archaeological implications, *Journal of World Prehistory*, 4, 157–222.

Hodder, I. (ed.) (1997) *On the surface: Çatalhöyük 1993–95.* Cambridge: McDonald Institute for Archaeological Research/British Institute of Archaeology at Ankara Monograph no. 22.

Hodell, D.A., Curtis, J.H. and Brenner, M. (1995) Possible role of climate in the collapse of Classic Maya civilisation, *Nature*, 375, 391–3.

Hodell, D.A., Curtis, J.H., Jones, G.A., Higuera-Gundy, A., Brenner, M., Binford, M.W. and Dorsey, K. (1991) Reconstruction of Caribbean climate change over the past 10 500 years, *Nature*, 52, 790–3.

Hofmann, W. (1986) Chironomid analysis, in B.E. Berglund (ed.), *Handbook of Holocene palaeoecology and palaeohydrology.* Chichester: John Wiley, pp. 715–28.

Hoganson, J.W. and Ashworth, A.C. (1992) Fossil beetle evidence for climatic change 18 000–10 000 years BP in south central Chile, *Quaternary Research*, 37, 101–16.

Holmes, J.A. (1992) Nonmarine ostracods as Quaternary palaeoenvironmental indicators, *Progress in Physical Geography*, 16, 405–31.

Hooghiemstra, H., Melice, J.L., Berger, A. and van der Hammen, T.
(1993) Frequency spectra and paleoclimatic variability of the high-
resolution 30–1450 Ka Funza I pollen record (Eastern Cordillera,
Colombia), *Quaternary Science Reviews*, 12, 141–56.

Hooke, J.M. and Kain, R.J.P. (1982) *Historical change in the physical
environment*. London: Butterworth.

Horowitz, A. and Gat, J.R. (1984) Floral and isotopic indications for
possible summer rains in Israel during wetter times, *Pollen et Spores*,
26, 61–8.

Hoskins, W.G. (1955) *The making of the English landscape*. London:
Hodder and Stoughton.

Hulme, M. (1994) Historic records and recent climatic change, in
N. Roberts (ed.), *The changing global environment*. Oxford: Blackwell, pp. 69–98.

Huntington, E. (1915) *Civilisation and climate*. New Haven: Yale
University Press.

Huntley, B. (1993a) Species-richness in north-temperate zone forests,
Journal of Biogeography, 20, 163–80.

Huntley, B. (1993b) Rapid early-Holocene migration and high abundance of hazel (*Corylus avellana* L.): alternative hypotheses, in
F.M. Chambers (ed.), *Climatic change and human impact on the
landscape*. London: Chapman and Hall, pp. 205–15.

Huntley, B. and Birks, H.J.B. (1983) *An atlas of past and present
pollen maps for Europe: 0–13 000 years ago*. Cambridge: Cambridge
University Press.

Huntley, B. and Prentice, I.C. (1988) July temperatures in Europe
from pollen data, 6000 years before present, *Science*, 241, 687–90.

Huntley, B. and Webb, T. (1989) Migration: species' response to
climatic variations caused by changes in the earth's orbit, *Journal
of Biogeography*, 16, 5–19.

Huntley, B., Bartlein, P.J. and Prentice, I.C. (1989) Climatic control of
the distribution and abundance of beech (*Fagus* L.) in Europe and
North America, *Journal of Biogeography*, 16, 551–60.

Hutchinson, G.E. (1957–75) *A treatise on limnology*, 3 vols. New
York: Wiley.

Hutchinson, G.E. (1973) Eutrophication, *American Scientist*, 61, 269–
79.

IUCN (1975) *World directory of national parks and other protected
areas*. Morges, Switzerland: IUCN.

Imbrie, J. and Imbrie, K.P. (1979) *Ice ages. Solving the mystery*.
London: Macmillan.

Innes, J.L. (1985) Lichenometry, *Progress in Physical Geography*, 9,
187–254.

Ionnides, M.G. (1937) *The regime of the rivers Euphrates and Tigris*.
London:

Islebe, G.A., Hooghiemstra, H. and Vanderborg, K. (1995) A cooling
event during the Younger Dryas Chron in Costa Rica, *Palaeogeography, Palaeoclimatology, Palaeoecology*, 117, 73–80.

Islebe, G.A., Hooghiemstra, H., Brenner, M., Curtis, J.H. and Hodell,
D.A. (1996) A Holocene vegetation history from lowland Guatemala, *The Holocene*, 6, 265–71.

Ivanovitch, M. and Harmon, R.S. (eds) (1995) *Uranium-series disequilibrium: applications to earth, marine and environmental sciences*, 2nd edn. Oxford: Clarendon Press.

Iversen, J. (1941) Landnám i Danmarks Stenalder (*Landnám* in Denmark's Stone Age), *Danmarks Geologiske Undersogelse*, II Rk. Nr 66, 1–68.

Iversen, J. (1944) *Viscum, Hedera and Ilex* as climate indicators. A contribution to the study of the post-glacial temperature climate. *Geologiska Föreningens i Stockholm Förhandlingar*, 66, 463–83.

Iversen, J. (1958) The bearing of glacial and interglacial epochs on the formation and extinction of plant taxa, *Uppsala Universiteit Arssk*, 6, 210–15.

Jacobi, R.M. (1978) Northern England in the eighth millennium BC: an essay, in P. Mellars (ed.), *The early postglacial settlement of northern Europe*. London: Duckworth, pp. 295–332.

Jacobi, R.M., Tallis, J.H. and Mellars, P.A. (1976) The southern Pennine Mesolithic and the ecological record, *Journal of Archaeological Science*, 3, 307–20.

Jacobsen, T. (1982) *Salinity and irrigation agriculture in antiquity*. Chicago: Bibliotheca Mesopotamica, vol. 14.

Jacobsen, T.W. and Farrand, W.R. (1987) *Excavations at Franchthi cave, Greece. Fascicle 1. General introduction to the excavations*. Bloomington, Indiana: Indiana University Press.

James, P.A. and Chester, D.K. (1995) Soils of Quaternary river sediments in the Algarve, in J. Lewin, M.G. Macklin and J.C. Woodward (eds), *Mediterranean Quaternary river environments*. Rotterdam: A.A. Balkema, pp. 245–62.

Jameson, M.H., Runnels, C.N. and van Andel, T.H. (1994) *A Greek countryside. The Southern Argolid from prehistory to the present day*. Stanford, California: Stanford University Press.

Jarman, M.R., Bailey, G.N. and Jarman, H.N. (eds) (1982) *Early European agriculture*. Cambridge: Cambridge University Press.

Jenkinson, R.D.S. (1984) *Cresswell Crags: late Pleistocene sites in the East Midlands*. Oxford: British Archaeological Reports, British Series 122.

Johnson, D.L. (1977) The late Quaternary climate of coastal California: evidence for an ice age refugium, *Quaternary Research*, 8, 154–79.

Jones, M. (1986) *England before Domesday*. London: Batsford.

Jones, R. (1977) Man as an element of a continental fauna: the case of the sundering of the Bassian bridge, in J. Allen, J. Golson and R. Jones (eds), *Sunda and Sahul. Prehistoric Studies in Southeast Asia, Melanesia and Australia*. London: Academic Press, pp. 317–85.

Jones, R. (1989) East of Wallace's line: issues and problems in the colonization of the Australian continent, in P. Mellars and C.B. Stringer (eds), *The human revolution*. Edinburgh: Edinburgh University Press, pp. 743–82.

Jones, V.J., Stevenson, A.C. and Battarbee, R.W. (1989) Acidification of lakes in Galloway, south west Scotland: a diatom and pollen study of the post-glacial history of the Round Loch of Glenhead, *Journal of Ecology*, 77, 1–23.

Jouzel, J., Barkov, N.I., Barnola, J.M., Bender, M., Chappellaz, J., Genthon, C., Kotlyakov, V.M., Lipenkov, V., Lorius, C., Petit, J.R., Raynaud, D., Raisbeck, G., Ritz, C., Sowers, T., Stievenard, M.,

Yiou, F. and Yiou, P. (1993) Extending the Vostok ice-core record of palaeoclimate to the penultimate glacial period, *Nature*, 364, 407–12.

Keeley, L.H. (1992) The introduction of agriculture to western North European Plain, in A.B. Gebauer and T.D. Price (eds), *Transitions to agriculture in prehistory*. Madison, Wisconsin: Prehistory Press, pp. 81–96.

Kelts, K. and Hsü, K.J. (1978) Freshwater carbonate sedimentation, in A. Lerman (ed.), *Lakes, chemistry, geology, physics*. New York: Springer-Verlag, pp. 295–323.

Kerney, M.P. (1977) A proposed zonation scheme for late-glacial and post-glacial deposits using land mollusca, *Journal of Archaeological Science*, 4, 387–90.

Kerney, M.P., Brown, E.H. and Chandler, T.J. (1963) The late-glacial and post-glacial history of the chalk escarpment near Brook, Kent, *Philosophical Transactions of the Royal Society of London, Series B*, 248, 135–204.

Kershaw, A.P. (1978) Record of last inter-glacial cycle from north-eastern Queensland, *Nature*, 272, 159–61.

Kershaw, A.P. (1986) Climatic change and Aboriginal burning in northeast Australia during the last two glacial/interglacial cycles, *Nature*, 322, 47–9.

Khazanov, A.M. (1984) *Nomads and the outside world*. Cambridge: Cambridge University Press.

Klein, R.G. (1980) Later Pleistocene hunters, in A. Sherratt (ed.), *The Cambridge Encyclopeadia of Archaeology*. Cambridge: Cambridge University Press, pp. 87–95.

Knox, J.C. (1984) Responses of river systems to Holocene climates, in H.E. Wright (ed.), *Late Quaternary environments of the United States. Volume 2, The Holocene*. London: Longman, pp. 26–41.

Knox, J.C. (1993) Large increases in flood magnitude in response to modest changes in climate, *Nature*, 301, 430–2.

Kochel, R.C. and Baker, V.R. (1982) Paleoflood hydrology, *Science*, 215, 353–61.

Köhler-Rollefson, I. (1996) The one-humped camel in Asia: origin, utilization and mechanisms of dispersal, in D.R. Harris (ed.), *The origins and spread of agriculture and pastoralism in Eurasia*. London: UCL Press, pp. 282–94.

Kosse, K. (1979) *Settlement ecology of the Koros and Linear Pottery cultures in Hungary*. Oxford: British Archaeological Reports, International Series 64.

Kraft, J.C., Aschenbrenner, S.E. and Rapp, G. (1977) Paleogeographic reconstructions of coastal Aegean archaeological sites, *Science*, 195, 41–7.

Kraft, J.C., Kayan, I. and Erol, O. (1980) Geomorphic reconstructions in the environs of ancient Troy, *Science*, 209, 776–82.

Krantz, G.S. (1970) Human activities and megafaunal extinctions, *American Scientist*, 58, 164–70.

Kreiser, A.M., Appleby, P.G., Natkanski, J., Rippey, B. and Battarbee, R.W. (1990) Afforestation and lake acidification: a comparison of four sites in Scotland, *Philosophical Transactions of the Royal Society of London, Series B*, 327, 377–83.

Kromer, B. and Becker, B. (1993) German oak and pine ^{14}C calibration 7200–9400 BC, *Radiocarbon*, 35, 125–37.

Krzyzaniak, L. (1991) Early farming in the Middle Nile basin: recent discoveries at Kadero (central Sudan), *Antiquity*, 65, 515–32.

Kudrass, H.R., Erlenkeuser, H., Vollbrecht, R. and Weiss, W. (1991) Global nature of the Younger Dryas cooling event inferred from oxygen isotope data from Sulu Sea cores, *Nature*, 349, 406–9.

Kumar, N., Anderson, R.F., Mortlock, R.A., Froelich, P.N., Kubik, P., Dittrich-Hannen, B. and Suter, M. (1995) Increased biological productivity and export production in the glacial Southern Ocean, *Nature*, 378, 675–80.

Kuniholm, P.I., Kromer, B., Manning, S.W., Newton, M., Latini, C.E. and Bruce, M.J. (1996) Anatolian tree rings and the absolute chronology of the eastern Mediterranean, 2220–718 BC, *Nature*, 381, 780–3.

Kurtén, B. and Anderson, E. (1980) *Pleistocene mammals of North America*. New York: Columbia University Press.

Küster, H. (1997) The role of farming in the postglacial expansion of beech and hornbeam in the oak woodlands of central Europe, *The Holocene*, 7, 239–42.

Kutzbach, J. (1983) Monsoon rains of the late Pleistocene and early Holocene: patterns, intensity and possible causes of changes, in A. Street-Perrott, M. Beran and R. Ratcliffe (eds), *Variations in the global water budget*. Dordrecht: Reidel, pp. 371–89.

Kutzbach, J. and Street-Perrott, F.A. (1985) Milankovitch forcing of fluctuations in the level of tropical lakes from 18 to 0 kyr BP, *Nature*, 317, 130–4.

Kutzbach, J.E. (1980) Estimates of past climate at Paleolake Chad, North Africa, based on a hydrological and energy balance model, *Quaternary Research*, 14, 210–23.

Kutzbach, J.E. and Gallimore, R.G. (1988) Sensitivity of a coupled atmosphere/mixed layer ocean model to changes in orbital forcing at 9000 years BP, *Journal of Geophysical Research*, 93, 803–21.

Kutzbach, J.E., Bonan, G., Foley, J.A. and Harrison, S.P. (1996) Vegetation and soil feedbacks on the response of the African monsoon to orbital forcing in the early to middle Holocene, *Nature*, 384, 623–6.

Ladurie, E. le Roy (1972) *Times of feast, times of famine*. London: George Allen and Unwin.

Laird, K.R., Fritz, S.C., Maasch, K.A. and Cumming, B.F. (1996) Greater drought intensity and frequency before AD 1200 in the Northern Great Plains, USA, *Nature*, 384, 552–4.

LaMarche, V.C. and Hirschboek, K.K. (1984) Frost rings in trees as records of major volcanic eruptions, *Nature*, 307, 121–6.

Lamb, H.F. and van der Kaars, S. (1995) Vegetational response to Holocene climatic change: pollen and palaeolimnological data from the Middle Atlas, Morocco, *The Holocene*, 5, 400–8.

Lamb, H.F., Damblon, F. and Maxted, R.W. (1991) Human impact on the vegetation of the Middle Atlas, Morocco, during the last 5000 years, *Journal of Biogeography*, 18, 519–32.

Lamb, H.F., Gasse, F., Benkaddour, A., el-Hamouti, N., van der Kaars, S., Perkins, W.T., Pearce, N.J. and Roberts, C.N. (1995) Relation between century-scale Holocene arid intervals in tropical and temperate zones, *Nature*, 373, 134–7.

Lamb, H.H. (1977) *Climate: present, past and future. Volume 2, Climatic history and the future*. London: Methuen.

Lamb, H.H. (1981) An approach to the study of the development of climate and its impact in human affairs, in T.M.L. Wigley, M.J. Ingram and G. Farmer (eds), *Climate and history*. Cambridge: Cambridge University Press, pp. 291–309.

Lamb, H.H. (1982) *Climate, history and the modern world*. London and New York: Methuen.

Lambert, R., Faulkner, R., Bell, M., Hotchkiss, P., Roberts, N. and Windram, A. (1990) The use of wetlands (dambos) for micro-scale irrigation in Zimbabwe, *Irrigation and Drainage Systems*, 4, 17–28.

Lancaster, N. (1990) Palaeoclimatic evidence from sand seas, *Palaeogeography, Palaeoclimatology, Palaeoecology*, 74, 279–90.

Langbein, W.B. (1961) Salinity and hydrology of closed lakes, *United States Geological Survey, Professional Paper 412*.

Larsen, C.P.S. (1996) Fire and climate dynamics in the boreal forest of northern Alberta, Canada, from AD 1850 to 1989, *The Holocene*, 6, 449–56.

LaSalle, P. and Shilts, W.M. (1993) Younger Dryas-age readvance of Laurentide ice into the Champlain Sea, *Boreas*, 22, 25–37.

Leavitt, P.R. (1993) A review of factors that regulate carotenoid and chlorophyll deposition and fossil pigment abundance, *Journal of Paleolimnology*, 9, 109–27.

Lee, R.B. and DeVore, I. (eds) (1968) *Man the hunter*. Chicago: Aldine.

Legge, A.J. (1996) The beginning of caprine domestication in Southwest Asia, in the Near East, in D.R. Harris (ed.), *The origins and spread of agriculture and pastoralism in Eurasia*. London: UCL Press, pp. 238–62.

Legge, A.J. and Rowley-Conwy, P.A. (1988) *Star Carr revisited*. London: University of London, Centre for Extra-Mural Studies.

Lehtonen, H. and Huttunen, P. (1997) History of forest fires in eastern Finland from the fifteenth century AD – the possible effects of slash-and-burn cultivation, *The Holocene*, 7, 223–8.

Lemcke, G. and Sturm, M. (1997) $\partial^{18}O$ and trace element measurements as proxy for the reconstitution of climate changes at Lake Van (Turkey): preliminary results, in H.N. Dalfez, G. Kukla and H. Weiss (eds), *Third millennium BC climate change and the Old World collapse*. Proceedings of NATO ASI Series I, Vol. 49, Springer-Verlag, pp. 653–78.

Lewin, R. (1993) *The origin of modern humans*. Scientific American/ W.H. Freeman.

Lewis, H.T. (1972) The role of fire in the domestication of plants and animals in southwest Asia: a hypothesis, *Man*, 7, 195–222.

Lewthwaite, J. (1989) Isolating the residuals: the mesolithic basis of man–animal relationships on the Mediterranean islands, in C. Bonsall (ed.), *The Mesolithic in Europe*. Edinburgh: John Donald, pp. 541–55.

Lewthwaite, J.W. and Sherratt, A. (1980) Chronological atlas, in A. Sherratt (ed.), *Cambridge Encyclopaedia of Archaeology*. Cambridge: Cambridge University Press, pp. 437–52.

Leyden, B.W. (1987) Man and climate in the Maya lowlands, *Quaternary Research*, 28, 407–15.

Lézine, A.-M. (1989) Late Quaternary vegetation and climate of the Sahel, *Quaternary Research*, 32, 317–34.

Limbrey, S. (1975) *Soil science and archaeology*. London: Academic Press.

Limbrey, S. (1990) Edaphic opportunism? A discussion of soil factors in relation to the beginnings of plant husbandry in south-west Asia, *World Archaeology*, 22, 45–52.

Livingstone, D.A. (1957) On the sigmoid growth phase in the history of Linsley Pond, *American Journal of Science*, 255, 364–73.

Lock, W.W., Andrews, J.T. and Webber, P.J. (1979) A manual for lichenometry, *B.G.R.G. Technical Bulletin*, 26.

Lorius, C., Jouzel, J., Raynaud, D., Hansen, J. and Le Treut, H. (1990) The ice-core record: climate sensitivity and future greenhouse warming, *Nature*, 347, 139–45.

Lotter, A. (1991) Absolute dating of the Late-glacial period in Switzerland using annually laminated lake sediments, *Quaternary Research*, 35, 321–30.

Lotter, A.F., Eicher, U., Birks, H.J.B. and Siegenthaler, U. (1992) Late-glacial climatic oscillations as recorded in Swiss lake sediments, *Journal of Quaternary Science*, 7, 187–204.

Lotter, A.F., Sturm, M., Teranes, J.L. and Wehrli, B. (1997) Varve formation since 1885 and high-resolution varve analyses in hypertrophic Baldeggersee (Switzerland), *Aquatic Science*, 59, 1–22.

Lowe, J.J. and Walker, M.J.C. (1997) *Reconstructing Quaternary environments*, 2nd edn. London: Longman.

Lowe, J.J., Ammann, B., Birks, H.H., Björck, S., Coope, G.R., Cwynar, L., de Beaulieu, J.-L., Mott, R.J., Peteet, D.M. and Walker, M.J.C. (1994) Climatic change in areas adjacent to the North Atlantic during the last glacial–interglacial transition (14–9 ka BP): a contribution to IGCP-253, *Journal of Quaternary Science*, 9, 185–98.

Lowe, J.J., Coope, G.R., Sheldrick, C., Harkness, D.D. and Walker, M.J.C. (1995) Direct comparison of UK temperatures and Greenland snow accumulation rates, 15 000–12 000 yr ago, *Journal of Quaternary Science*, 10, 175–80.

Ložek, V. (1986) Mollusca analysis, in B.E. Berglund (ed.), *Handbook of Holocene palaeoecology and palaeohydrology*. Chichester: John Wiley, pp. 729–40.

Lubell, D., Hassan, F.A., Gautier, A. and Ballais, J.-L. (1976) The Capsian Escargotières, *Science*, 191, 910–20.

Luly, J. (1993) Holocene palaeoenvironments near Lake Tyrrell, semi-arid northwestern Victoria, *Journal of Biogeography*, 20, 587–98.

McCarroll, D. (1995) A new approach to lichenometry: dating single-age and diachronous surfaces, *The Holocene*, 4, 383–96.

MacCarthy, M.R. (1995) Archaeological and environmental evidence for the Roman impact on vegetation near Carlisle, Cumbria, *The Holocene*, 5, 491–6.

McClure, H.A. (1976) Radiocarbon chronology of late Quaternary lakes in the Arabian desert, *Nature*, 263, 755–6.

McDonald, J.N. (1981) *North American bison: their classification and evolution*. Berkeley and San Francisco: University of California Press.

McDonald, J.N. (1984) The recorded North American selection regime and late Quaternary megafaunal estimations, in P.S. Martin and R.G. Klein (eds), *Quaternary extinctions*. Tucson: University of Arizona Press, pp. 404–39.

McGlone, M.S. (1983) Polynesian deforestation of New Zealand: a preliminary synthesis, *Archaeology in Oceania*, 18, 11–25.

McGlone, M.S., Anderson, A.J. and Holdaway, R.N. (1994) An ecological approach to the Polynesian settlement of New Zealand, in D.G. Sutton (ed.), *The origin of the first New Zealanders*. Auckland University Press, pp. 136–63.

McGlone, M.S., Salinger, M.J. and Moar, N.T. (1993) Palaeovegetation studies of New Zealand's climate since the last glacial maximum, in H.E. Wright, Jr, J.E. Kutzbach, T. Webb III, W.F. Ruddiman, F.A. Street-Perrott and P.J. Bartlein (eds), *Global climates since the last glacial maximum*. Minneapolis: University of Minnesota Press, pp. 294–317.

McGovern, T.H. (1980) Cows, harpseals and churchbells: adaptation and extinction in Norse Greenland, *Human Ecology*, 8, 245–77.

McGreevy, T. (1989) Prehistoric pastoralism in northern Peru, in J. Clutton-Brock (ed.), *The walking larder. Patterns of domestication, pastoralism and predation*. London: Unwin Hyman, pp. 231–9.

McIntosh, R.J. (1983) Floodplain geomorphology and human occupation of the upper inland delta of the Niger, *Geographical Journal*, 149, 182–201.

Mackereth, F.J.H. (1966) Some chemical observations on post-glacial lake sediments, *Philosophical Transactions of the Royal Society of London, Series B*, 250, 165–213.

Macklin, M.G., Rumsby, B.T. and Heap, T. (1992) Flood alluviation and entrenchment: Holocene valley-floor development and transformation in the British uplands, *Bulletin of the Geological Society of America*, 104, 631–43.

McNeil, J.R. (1992) *The mountains of the Mediterranean world: an environmental history*. Cambridge: Cambridge University Press.

MacNeish, R.S. (1967) A summary of the subsistence, in D.S. Bayers (ed.), *Environment and subsistence: the prehistory of the Tehuacan valley, Vol. 1*. Austin: University of Texas Press, pp. 290–309.

Magri, D. (1995) Some questions on the late-Holocene vegetation of Europe, *The Holocene*, 5, 354–60.

Magri, D. (1997) Middle and late Holocene vegetation and climate changes in Peninsular Italy, in H.N. Dalfez, G. Kukla and H. Weiss (eds), *Third millennium BC climate change and the Old World collapse*. Proceedings of NATO ASI Series I, Vol. 49, Springer-Verlag, pp. 517–30.

Maguire, D.J. (1983) The identification of agricultural activity using pollen analysis, in M. Jones (ed.), *Integrating the subsistence economy*. Oxford: British Archaeological Reports, International Series 181, pp. 5–18.

Maizels, J. and Aitken, J. (1991) Palaeohydrological change during deglaciation in upland Britain: a case study from northeast Scotland, in L. Starkel, K.J. Gregory and J.B. Thornes (eds), *Temperate palaeohydrology. Fluvial processes in the temperate zone during the last 15 000 years*. Chichester: John Wiley, pp. 105–45.

Maley, J. (1981) *Etudes palynologiques dans le bassin du Tchad et paléoclimatologie de l'Afrique nord-tropicale de 30 000 ans à l'époque actuelle.* Paris: ORSTOM travaux et documents 129.

Maley, J. and Livingstone, D.A. (1983) Extension d'un élément montagnard dans le sud du Ghana (Afrique de l'Ouest) au Pléistocène supérieur et à l'Holocène inférieur: premières données polliniques, *Comptes rendus de l'Académie des Sciences*, série 3, 296, 251–6.

Maloney, B.K. (1993) Palaeoecology and the origins of the coconut, *GeoJournal*, 31, 355–62.

Mandel, W.M. (1969) Soviet marxism and the social science, in A. Simirenko (ed.), *Social thought in the Soviet Union.* Chicago: Quadrangle Press.

Mangelsdorf, P.C. (1974) *Corn: its origin, evolution and improvement.* Cambridge, Massachussets: Harvard University Press.

Mangerud, J., Birks, H.J.B. and Jager, K.D. (eds) (1982) Chronostratigraphic subdivision of the Holocene, *Striae*, 16.

Markgraf, V. (1993) Climatic history of Central and South America since 18 000 yr BP: comparison of pollen records and model simulations, in H.E. Wright, Jr, J.E. Kutzbach, T. Webb III, W.F. Ruddiman, F.A. Street-Perrott and P.J. Bartlein (eds), *Global climates since the last glacial maximum.* Minneapolis: University of Minnesota Press. pp. 357–85.

Martienssen, R. (1997) The origin of maize branches out, *Nature*, 386, 443–4.

Martin, J.H. and Fitzwater, S.E. (1988) Iron deficiency limits phytoplankton growth in the north-east Pacific subarctic, *Nature*, 331, 341–3.

Martin, P.S. and Klein, R.G. (eds) (1984) *Quaternary extinctions.* Tucson: University of Arizona Press.

Matthews, J.A. (1985) Radiocarbon dating of surface and buried soils: principles, problems and prospects, in K.S. Richards, R.R. Arnett and S. Ellis (eds), *Geomorphology and soils.* London: Allen and Unwin, pp. 271–88.

Matthews, J.A. (1992) *The ecology of recently deglaciated terrain.* Cambridge: Cambridge University Press.

Mayewski, P.A., Meeker, L.D., Whitlow, S., Twickler, M.S., Morrison, M.C., Grootes, P.M., Bond, G.C., Alley, R.B., Meese, D.A., Gow, A.J., Taylor, K.C., Ram, M. and Wumkes, M. (1994) Changes in atmospheric circulation and ocean ice cover over the North Atlantic during the last 41 000 years, *Science*, 263, 1747–51.

Mayewski, P.A., Twickler, M.S., Whitlow, S., Meeker, L.D., Yang, Q., Thomas, J., Kreutz, K., Grootes, P.M., Morse, D.L., Steig, E.J., Waddington, E.D., Saltzman, E.S., Whung, P.-Y. and Taylor, K.C. (1994) Climate change during the last deglaciation in Antarctica, *Science*, 272, 1636–8.

Mayle, F.E. and Cwynar, L.C. (1995) A review of multi-proxy data for the Younger Dryas in Atlantic Canada, *Quaternary Science Reviews*, 14, 813–21.

Meadow, R.H. (1989) Osteological evidence for the process of animal domestication, in J. Clutton-Brock (ed.), *The walking larder. Patterns of domestication, pastoralism and predation.* London: Unwin Hyman, pp. 80–90.

Meese, D.A., Gow, A.J., Grootes, P.M., Mayewski, P.A., Ram, M., Stuiver, M., Taylor, K.C., Waddington, E.D. and Zielinski, C.A. (1994) An accumulation record from the GISP2 core as an indicator of climate change throughout the Holocene, *Science*, 266, 1680–2.

Mellaart, J. (1967) *Çatal Hüyük: a neolithic town in Anatolia*. London: Thames and Hudson.

Mellars, P. (1975) Fire ecology, animal populations and man: a study of some ecological relationships in prehistory, *Proceedings of the Prehistoric Society*, 42, 15–45.

Mellars, P. (1976) Settlement patterns and industrial variability in the British Mesolithic, in G. de G. Sieveking, I.H. Longworth and K.E. Wilson (eds), *Problems in economic and social archaeology*. London: Duckworth, pp. 375–99.

Metcalfe, S.E., Street-Perrott, F.A., Brown, R.B., Hales, P.E., Perrott, R.A. and Steininger, F.M. (1989) Late Holocene human impact on lake basins in central Mexico, *GeoArchaeology*, 4, 119–41.

Miller, N.F. (1991) The Near East, in W. van Zeist, K. Wasylikowa and K.-E. Behre (eds), *Progress in Old World palaeoethnobotany*. Rotterdam: A.A. Balkema, pp. 133–60.

Millspaugh, S.H. and Whitlock, C. (1995) A 750-yr fire history based on lake sediment records in central Yellowstone National Park, *The Holocene*, 5, 283–92.

Mitchell, F.J.G. (1990) The impact of grazing and human disturbance on the dynamics of woodland in S.W. Ireland, *Journal of Vegetation Science*, 1, 245–54.

Mitchell, J.F.B. (1990) Greenhouse warming: is the Holocene a good analogue? *Journal of Climate*, 3, 1177–92.

Moore, A.M.T. (1985) The development of neolithic societies in the Near East, in F. Wendorf and A.E. Close (eds), *Advances in world archaeology*, Vol. 4. London: Academic Press, pp. 1–70.

Moore, A.M.T. and Hillman, G.C. (1992) The Pleistocene to Holocene transition and human economy in Southwest Asia: the impact of the Younger Dryas, *American Antiquity*, 57, 482–94.

Moore, P.D. (1975) Origin of blanket mires, *Nature*, 256, 267–9.

Moore, P.D., Merryfield, D.L. and Price, M.D.R. (1984) The vegetation and development of blanket mires, in P.D. Moore (ed.), *European mires*. London: Academic Press, pp. 203–35.

Moore, P.D., Webb, J.A. and Collinson, M.E. (1991) *An illustrated guide to pollen analysis*, 2nd edn. Oxford: Blackwell Scientific.

Mörner, N.-A. (ed.) (1976) The Pleistocene/Holocene boundary: a proposed boundary-stratotype in Gothenberg, Sweden, *Boreas*, 5, 193–275.

Moser, K.A., MacDonald, G.M. and Smol, J.P. (1996) Applications of freshwater diatoms to geographical research, *Progress in Physical Geography*, 20, 21–52.

Moss, B. (1980) *The ecology of freshwaters*. Oxford: Blackwell.

Mott, R.J., Grant, D., Stea, R. and Ochietti, S. (1986) Late-glacial climatic oscillation in Atlantic Canada equivalent to the Allerod/Younger Dryas event, *Nature*, 323, 247–50.

Murozumi, M., Chow, T.J. and Patterson, C. (1969) Chemical concentrations of pollutant lead aerosols, terrestrial dusts and sea salt in

Greenland and Antarctic snow strata, *Geochimica et Cosmochimica Acta*, 33, 1247–94.

Myers, N. (1979) *The sinking ark*. Oxford: Pergamon.

Naval Intelligence Division (1945) *Geographical Handbook Series, Spain and Portugal. Volume IV, The Atlantic Islands*.

Nelson, D.J., Webb, R.H. and Long, A. (1990) Analysis of stick-nest rat (*Leporillus*: Muridae) middens from central Australia, in J.L. Betancourt, T.R. van Devender and P.S. Martin (eds), *Packrat middens. The last 40 000 years of biotic change*. Tucson: University of Arizona Press, pp. 428–34.

Nesje, A. and Johannessen, T. (1992) What were the primary forcing mechanisms of high-frequency Holocene climate and glacier variations?, *The Holocene*, 2, 79–84.

Nicholson, S.E. (1980) Saharan climates in historic times, in H. Faure and M.E.J. Williams (eds), *The Sahara and the Nile*. Rotterdam: A.A. Balkema, pp. 173–200.

Nicholson, S.E. and Flohn, H. (1980) African environmental and climatic changes and the general atmospheric circulation in late Pleistocene and Holocene, *Climatic Change*, 2, 313–48.

Nicol-Pichard, S. (1987) Analyse pollinique d'une séquence tardi et postglaciaire à Tourves (Var, France), *Ecologia Mediterranea*, 12, 29–42.

Niederberger, C. (1979) Early sedentary economy in the basin of Mexico, *Science*, 203, 131–42.

Nunn, P.D. (1990) Coastal processes and landforms of Fiji: their bearing on Holocene sea-level changes in the south and west Pacific, *Journal of Coastal Research*, 6, 279–310.

O'Connor, J.E. and Baker, V.R. (1992) Magnitude and implications of peak discharges from glacial Lake Missoula, *Bulletin of the Geological Society of America*, 104, 267–79.

O'Hara, S.L., Street-Perrott, F.A. and Burt, T.P. (1993) Accelerated soil erosion around a Mexican highland lake caused by prehistoric agriculture, *Nature*, 362, 48–51.

O'Sullivan, P.E. (1979) The ecosystem watershed concept in the environmental sciences – a review, *International Journal of Environmental Sciences*, 13, 273–81.

O'Sullivan, P.E. (1983) Annually-laminated lake sediments and the study of Quaternary environmental changes – a review, *Quaternary Science Reviews*, 1, 245–313.

O'Sullivan, P.E. (1992) The eutrophication of shallow coastal lakes in Southwest England – understanding and recommendations for restoration, based on palaeolimnology, historical records, and the modelling of changing phosphorus loads, *Hydrobiologia*, 243/244, 421–34.

Oates, D. and Oates, J. (1976) Early irrigation agriculture in Mesopotamia, in G. de G. Sieveking, I.H. Longworth and K.E. Wilson (eds), *Problems in economic and social archaeology*. London: Duckworth, pp. 109–36.

Ogilvie, A.E.J. (1992) Documentary evidence for changes in the climate of Iceland, AD 1500 to 1800, in R.S. Bradley and P.D. Jones (eds), *Climate since AD 1500*. London: Routledge, pp. 92–117.

Oldfield, F. (1977) Lakes and their drainage basins as units of sediment-based ecological study, *Progress in Physical Geography*, 1, 460–504.

Oldfield, F. (1983) Man's impact on the environment: some recent perspectives, *Geography*, 68, 245–56.

Oldfield, F. and Appleby, P.G. (1984) Empirical testing of ^{210}Pb-dating models for lake sediments, in E.Y. Haworth and J.W.G. Lund (eds), *Lake sediments and environmental history*. Leicester: Leicester University Press, pp. 3–124.

Oldfield, F. and Clark, R.L. (1990) Lake sediment-based studies of soil erosion, in J. Boardman, J.A. Dearing and I.D.L. Foster (eds), *Soil erosion on agricultural land*. Chichester: John Wiley, pp. 201–30.

Oldfield, F., Tolonen, K. and Thompson, R. (1981) History of particulate atmospheric pollution from magnetic measurements in dated Finnish peat profiles, *Ambio*, 10, 185–8.

Oldfield, F., Worsley, A.T. and Appleby, P.G. (1985) Evidence from lake sediments for recent erosion rates in the highlands of Papua New Guinea, in I. Douglas and T. Spencer (eds), *Environmental change and tropical geomorphology*. London: George Allen and Unwin, pp. 185–96.

Olsson, I.U. (1986) Radiometric dating, in B.E. Berglund (ed.), *Handbook of Holocene palaeoecology and palaeohydrology*. Chichester: John Wiley, pp. 273–312.

Owen, R. (1849) On Dinornis, an extinct Genus of tridactyle struthious birds, with descriptions of portions of the skeleton of five species which formerly existed in New Zealand, *Transactions of the Zoological Society of London*, 3, 235–75.

Owen, R.B., Crossley, R., Johnson, T.C., Tweddle, D., Kornfield, I., Davison, S., Eccles, D.H. and Engstrom, D.E. (1990) Major low levels of Lake Malawi and their implications for speciation rates in cichlid fishes, *Proceedings of the Royal Society of London, Series B*, 240, 519–53.

Parry, M.L. (1978) *Climatic change, agriculture and settlement*. Folkestone: Dawson.

Patrick, S.T. (1988) Septic tanks as sources of phosphorus to Lough Erne, Ireland, *Journal of Environmental Management*, 26, 239–48.

Patterson, W.A. and Backman, A.E. (1988) Fire and disease history of forests, in B. Huntley and T. Webb III (eds), *Vegetation history*. Dordrecht: Kluwer, pp. 603–32.

Patterson, W.A., Edwards, K.J. and Maguire, D.J. (1987) Microscopic charcoal as a fossil indicator of fire, *Quaternary Science Reviews*, 6, 3–24.

Payne, S. (1975) Faunal change at Franchthi cave from 20 000 BC to 3 000 BC, in A.T. Clason (ed.), *Archaeo-zoological Studies*. Amsterdam: Elsevier, pp. 120–31.

Peglar, S. (1993) The mid-Holocene *Ulmus* decline at Diss Mere, Norfolk, UK: a year-by-year pollen stratigraphy from annual laminations, *The Holocene*, 3, 1–13.

Peglar, S. and Birks, H.J.B. (1993) The mid-Holocene *Ulmus* fall at Diss Mere, south-east England – disease and human impact?, *Vegetational history and archaeobotany*, 2, 61–8.

Pennington, W. (1977) The late Devensian flora and vegetation of the British Isles, *Philosophical Transactions of the Royal Society of London, Series B*, 280, 247–71.

Pennington, W. (1986) Lags in adjustment of vegetation to climate caused by the pace of soil development: evidence from Britain, *Vegetation*, 67, 105–18.

Pennington, W. and Bonny, A.P. (1970) Absolute pollen diagram from the British late-glacial, *Nature*, 226, 871–3.

Pennington, W., Cambray, R.S. and Fisher, E.M. (1973) Observations of lake sediment using fallout Cs-137 as a tracer, *Nature*, 242, 324–6.

Pepper, D. (1996) *Modern environmentalism.* London: Routledge.

Perrot, J. (1966) Le gisement Natoufien de Mallaha (Eynan), Israel, *L'Anthropologie*, 70, 437–83.

Perry, I. and Moore, P.D. (1987) Dutch elm disease as an analogue of Neolithic elm decline, *Nature*, 326, 72–3.

Peteet, D. (ed.) (1993) Global Younger Dryas?, *Quaternary Science Reviews*, 12 (5), 277–355.

Petit-Maire, N. and Riser, J. (eds) (1983) *Sahara ou Sahel? Quaternaire récent du bassin de Taoudenni (Mali).* Paris: Librairie du Muséum.

Pickersgill, B. (1989) Cytological and genetical evidence on the domestication and diffusion of crops within the Americas, in D.R. Harris and G.C. Hillman (eds), *Farming and foraging: the evolution of plant exploitation.* London: Unwin Hyman, pp. 426–39.

Pickersgill, B. and Heiser, C.B. (1977) Origin and distribution of plants domesticated in the New World tropics, in C.A. Reed (ed.), *Origins of agriculture.* The Hague: Mouton, pp. 803–35.

Pilcher, J.R., Baillie, M.G.L., Schmidt, B. and Becker, B. (1984) A 7272-year tree-ring chronology for western Europe, *Nature,* 312, 150–2.

Pilcher, J.R., Smith, A.G., Pearson, G.W. and Crowder, A. (1971) Land clearance in the Irish neolithic: new evidence and interpretation, *Science*, 172, 560–2.

Pons, A. and Quezel, P. (1985) The history of the flora and vegetation and past and present human disturbance in the Mediterranean region, in C. Gomez-Campo (ed.), *Plant conservation in the Mediterranean area.* Dordrecht: Dr W. Junk, pp. 25–43.

Pons, A. and Reille, M. (1986) Nouvelles recherches pollenanalytiques à Padul (Granada): la fin du dernier glaciaire et l'Holocène, in F. Lopez-Vera (ed.), *Quaternary climate in western Mediterranean.* Madrid: pp. 405–20.

Pons, A., Toni, C. and Triat, H. (1979) Edification de la Camargue et histoire holocène de sa végétation, *Terre Vie, Rec. Ecol.*, suppl. 2, 13–30.

Porter, S.C. (1981) Glaciological evidence of Holocene climatic change, in T.M.L. Wigley, M.J. Ingram and G. Farmer (eds), *Climate and history.* Cambridge: Cambridge University Press, pp. 82–110.

Porter, S.C. and Zhisheng, A. (1995) Correlation between climate events in the North Atlantic and China during the last glaciation, *Nature*, 375, 305–8.

Prance, G.T. (ed.) (1982) *Biological diversification in the tropics.* New York: Columbia University Press.

Preece, R.C. (1993) Late Glacial and Post-Glacial molluscan successions from the site of the Channel Tunnel in SE England, *Scripta Geologica*, 2, 387–95.

Prentice, I.C., Guiot, J., Huntley, B., Jolly, D. and Cheddadi, R. (1996) Reconstructing biomes from palaeoecological data: a general method and its application to European pollen data at 0 and 6 ka, *Climate Dynamics*, 12, 185–94.

Prentice, I.C., Sykes, M.T., Lautenschlager, M., Harrison, S.P., Denisseanko, O. and Bartlein, P.J. (1993) Modelling global vegetation patterns and terrestrial carbon storage at the last glacial maximum, *Global Ecology and Biogeography Letters*, 3, 67–76.

Price, D.G. (1993) Dartmoor: the pattern of prehistoric settlement sites, *Geographical Journal*, 159, 261–80.

Provensal, M. (1995) The role of climate in landscape morphogenesis since the Bronze Age in Provence, southeastern France, *The Holocene*, 5, 348–53.

Punning, J.-M. (ed.) (1994) *The influence of natural and anthropogenic factors on the development of landscapes. The results of a comprehensive study in NE Estonia.* Talinn: Institute of Ecology, Estonian Academy of Sciences.

Quaini, M. (1982) *Geography and marxism.* Oxford: Blackwell.

Rackham, O. (1980) *Ancient woodland.* London: Edward Arnold.

Rackham, O. (1982) Land-use and the native vegetation of Greece, in M. Bell and S. Limbrey (eds), *Archaeological aspects of woodland ecology.* Oxford: British Archaeological Reports, International Series 146, pp. 177–98.

Rahmsdorf, S. (1995) Bifurcations of the Atlantic thermohaline circulation in response to changes in the hydrological cycle, *Nature*, 378, 145–9.

Raikes, R. (1967) *Water, weather and prehistory.* London: John Baker.

Ramsay, C.B. (1995) OxCal program v2.18. Oxford University Radiocarbon Accelerator Unit.

Rapp, A., Murray-Rust, D.H., Christiansson, C. and Berry, L. (1972) Soil erosion and sedimentation in four catchments near Dodoma, Tanzania, *Geografiska Annaler*, 54A, 255–318.

Rapp, G. and Gifford, J.A. (1982) Archaeological geology, *American Scientist*, 70, 45–53.

Redclift, M. (1984) *Development and the environmental crisis: red or green alternatives?* London: Methuen.

Reed, C.A. (1970) Extinction of mammalian megafauna in the Old World late Quaternary, *BioScience*, 20, 284–8.

Reid, M.A., Tibby, J.C., Penny, D. and Gell, P.A. (1995) The use of diatoms to assess past and present water quality, *Australian Journal of Ecology*, 20, 57–64.

Reille, M. and Pons, A. (1992) The ecological significance of sclerophyllous oak forests in the western part of the Mediterranean basin: a note on pollen analytical data, *Vegetation*, 99–100, 13–17.

Reille, M., Andrieu, V. and de Bealieu, J.-L. (1996) Les grands traits de l'histoire de la végétation des montagnes Méditerranéennes Occidentales, *Écologie*, 27, 153–69.

Renberg, I. (1990) A 12 000 year perspective of the acidification of Lilla Öresjön, southwest Sweden, *Philosophical Transactions of the Royal Society of London, Series B*, 327, 357–61.

Renberg, I. and Hellberg, T. (1982) The pH history of lakes in south-western Sweden as calculated from the subfossil diatom flora of the sediments, *Ambio*, 1, 30–3.

Renberg, I., Wik Persson, M. and Emteryd, O. (1994) Pre-industrial atmospheric lead contamination detected in Swedish lake sediments, *Nature*, 368, 323–6.

Renfrew, C. (1973) *Before civilization. The radiocarbon revolution and prehistoric Europe.* London: Jonathan Cape.

Richards, G. (1985) Fossil Mediterranean molluscs as sea-level indicators, *Geological Magazine*, 122, 373–81.

Richards, P. (1985) *Indigenous agricultural revolution.* London: Hutchinson.

Rindos, D. (1984) *The origins of agriculture. An evolutionary perspective.* London: Academic Press.

Ritchie, J.C. (1984) *Past and present vegetation of the far north-west of Canada.* Toronto: University of Toronto Press.

Ritchie, J.C. (1985) Late-Quaternary climatic and vegetational change in the lower Mackenzie basin, northwest Canada, *Ecology*, 66, 612–21.

Ritchie, J.C. (1987) *Postglacial vegetation of Canada.* Cambridge: Cambridge University Press.

Ritchie, J.C. (1995) Current trends in studies of long-term plant community dynamics, *New Phytologist*, 130, 469–94.

Ritchie, J.C. and Haynes, C.V. (1987) Holocene vegetation zonation in the eastern Sahara, *Nature*, 330, 645–7.

Ritchie, J.C., Eyles, C.H. and Haynes, C.V. (1985) Sediment and pollen evidence for an early to mid Holocene humid period in the eastern Sahara, *Nature*, 314, 352–5.

Roberts, L. (1989) Disease and death in the New World, *Science*, 246, 1245–7.

Roberts, N. (1983) Age, palaeoenvironments and climatic significance of late Pleistocene Konya lake, Turkey, *Quaternary Research*, 19, 154–71.

Roberts, N. (1984) Pleistocene environments in time and space, in R. Foley (ed.), *Community ecology and human adaptation in the Pleistocene.* London: Academic Press, pp. 25–53.

Roberts, N. (1991) Late Quaternary geomorphological change and the origins of agriculture in south central Turkey, *GeoArchaeology*, 6, 1–26.

Roberts, N. and Barker, P. (1993) Landscape stability and bio-geomorphic response to past and future climatic shifts in Africa, in D. Thomas and R. Allison (eds), *Landscape sensitivity.* Chichester: John Wiley, pp. 65–82.

Roberts, N. and Wright, Jr, H.E. (1993) Vegetational, lake-level and climatic history of the Near East and Southwest Asia, in H.E. Wright, Jr, J.E. Kutzbach, T. Webb III, W.F. Ruddiman, F.A. Street-Perrott and P.J. Bartlein (eds), *Global climates since the last glacial maximum.* Minneapolis: University of Minnesota Press, pp. 194–220.

Roberts, N., Boyer, P. and Parish, R. (1997a) Preliminary results of geoarchaeological investigations at Çatalhöyük, in I. Hodder (ed.), *On the surface: Çatalhöyük 1993–95.* Cambridge: McDonald Institute for Archaeological Research/British Institute of Archaeology at Ankara Monograph no. 22, pp. 19–40.

Roberts, N., Eastwood, W.J., Lamb, H.F. and Tibby, J.C. (1997b) The age and causes of mid-Holocene environmental change in southwest Turkey, in H.N. Dalfez, G. Kukla and H. Weiss (eds), *Third millennium BC climate change and the Old World collapse*. Proceedings of NATO ASI Series I, Vol. 49, Springer-Verlag, pp. 409–29.

Roberts, N., Taieb, M., Barker, P., Damnati, B., Icole, M. and Williamson, D. (1993) Timing of Younger Dryas climatic event in East Africa from lake-level changes, *Nature*, 366, 146–8.

Robertshaw, P. (1989) The development of pastoralism in East Africa, in J. Clutton-Brock (ed.), *The walking larder. Patterns of domestication, pastoralism and predation*. London: Unwin Hyman, pp. 207–14.

Rognon, P. (1983) Essai de définition et typologie des crises climatiques, *Bulletin de l'Institut de Géologie du bassin d'Aquitaine*, 34, 151–64.

Rose, J., Turner, C., Coope, G.R. and Bryan, M.D. (1980) Channel changes in a lowland river catchment over the last 13 000 years, in R.A. Cullingford, D.A. Davidson and J. Lewin (eds), *Timescales in geomorphology*. Chichester: John Wiley, pp. 159–75.

Rosenquist, I.T. (1978) Alternative sources for acidification of river water in Norway, *Science of the Total Environment*, 10, 39–49.

Rosman, K.J.R., Chisholm, W., Boutron, C.F., Candelone, J.P. and Görlach, U. (1993) Isotopic evidence for the source of lead in Greenland snows since the late 1960s, *Nature*, 362, 333–5.

Rossignol-Strick, M. (1995) Sea–land correlation of pollen records in the eastern Mediterranean for the glacial–interglacial transition: biostratigraphy versus radiometric time-scale, *Quaternary Science Reviews*, 14, 893–915.

Rothlisberger, F. and Schneebeli, W. (1979) Genesis of lateral moraine complexes, demonstrated by fossil soils and trunks; indicators of postglacial climatic fluctuations, in Ch. Schlüchter (ed.), *Moraines and varves*. Rotterdam: Balkema, pp. 387–420.

Rotnicki, K. (1991) Retrodiction of palaeodischarges of meandering and sinuous alluvial rivers and its palaeohydroclimatic implications, in L. Starkel, K.J. Gregory and J.B. Thornes (eds), *Temperate palaeohydrology. Fluvial processes in the temperate zone during the last 15 000 years*. Chichester: John Wiley, pp. 431–71.

Ruddiman, W.F. and McIntyre, A. (1981) The mode and mechanism of the last deglaciation: oceanic evidence, *Quaternary Research*, 16, 125–34.

Russell, J.W. (1989) *Modes of production in world history*. London: Routledge.

Ryan, W.B.F., Pitman III, W.C., Major, C.O., Shimkus, K., Moslalenko, V., Jones, G.A., Dimitrov, P., Görür, N., Sakinç, M. and Yüce, H. (1997) An abrupt drowning of the Black sea shelf, *Marine Geology*, 138, 119–26.

Rymer, L. (1980) Epistemology of historical ecology, *Environmental Conservation*, 7, 136–8.

Sadler, J.P. and Jones, J.C. (1997) Chironomids as indicators of Holocene environmental change in Britain, *Quaternary Proceedings*, 5, 1–16.

Sadler, J.P. and Skidmore, P. (1995) Introductions, extinctions or continuity? Faunal change in the North Atlantic islands, in R.A. Butlin and N. Roberts (eds), *Ecological relations in historical times: human impact and adaptation*. Oxford: Blackwell, pp. 206–25.

Sahlins, M. (1974) *Stone age economics*. London: Tavistock.

Sarnthein, M. (1978) Sand deserts during glacial maximum and climatic optimum, *Nature*, 272, 43–5.

Sauer, C.O. (1948) Environment and culture during the last deglaciation, *Proceedings of the American Philosophical Society*, 92, 65–77.

Sauer, C.O. (1952) *Agricultural origins and dispersals*. New York: American Geographical Society.

Schule, W. (1993) Mammals, vegetation and the initial human settlement of the Mediterranean islands: a palaeoecological approach, *Journal of Biogeography*, 20, 399–412.

Schumm, S.A. (1969) River metamorphosis, *Proceedings of the American Society of Civil Engineers, Journal of the Hydraulics Division*, 5, 255–73.

Schweingruber, F.H. (1988) *Tree rings*. Dordrecht: Kluwer.

Scott, L. (1990) Hyrax (Procaviidae) and dassie rat (Petromuridae) middens in paleoenvironmental studies in Africa, in J.L. Betancourt, T.R. van Devender and P.S. Martin (eds), *Packrat middens. The last 40 000 years of biotic change*. Tucson: University of Arizona Press, pp. 398–407.

Shackleton, J.C. (1988) Reconstructing past shorelines as an approach to determining factors affecting shellfish collecting in the prehistoric past, in G. Bailey and J. Parkington (eds), *The archaeology of prehistoric coastlines*. Cambridge: Cambridge University Press, pp. 11–21.

Shackleton, J.C. and van Andel, T.H. (1980) Prehistoric shell assemblages from Franchthi cave and evolution of the adjacent coastal zone, *Nature*, 288, 357–9.

Shaw, T. (1977) Early crops in Africa: a review of the evidence, in J.R. Harlan, J.M.J. de Wet and A.B.L. Stemler (eds), *Origins of African plant domestication*. The Hague: Mouton, pp. 107–54.

Shaw, T. (1980) Agricultural origins in Africa, in A. Sherratt (ed.), *The Cambridge Encyclopaedia of Archaeology*. Cambridge: Cambridge University Press, pp. 179–84.

Sheail, J. (1980) *Historical ecology: the documentary evidence*. Cambridge: Institute of Terrestrial Ecology.

Sherratt, A. (1980) Water, soil and seasonality in early cereal cultivation, *World Archaeology*, 11, 313–30.

Sherratt, A. (1981) Plough and pastoralism: aspects of the secondary products revolution, in I. Hodder, G. Isaac and N. Hammond (eds), *Pattern of the past. Studies in honour of David Clarke*. Cambridge: Cambridge University Press, pp. 261–305.

Shulmeister, J. (1992) A Holocene pollen record from lowland tropical Australia, *The Holocene*, 2, 107–16.

Simmons, A. (1991) Humans, island colonization and Pleistocene extinctions in the Mediterranean: the view from Akrotiri *Aetokremnos*, Cyprus, *Antiquity*, 65, 857–69.

Simmons, I. and Tooley, M. (eds) (1981) *The environment in British prehistory*. London: Duckworth.

Simmons, I.G. (1993) *Environmental history. A concise introduction.* Oxford: Blackwell.

Simmons, I.G. (1996) *Changing the face of the earth*, 2nd edn. Oxford: Blackwell.

Simmons, I.G. and Dimbleby, G.W. (1974) The possible role of ivy (Hedera helix L.) in the mesolithic economy of western Europe, *Journal of Archaeological Science*, 1, 291–6.

Simmons, I.G. and Innes, J.B. (1985) Late Mesolithic land-use and its impact in the English uplands, in R.T. Smith (ed.), *The biogeographical impact of land-use change: collected essays.* Biogeographical Monographs 2, pp. 7–17.

Simmons, I.G. and Innes, J.B. (1996) The ecology of an episode of prehistoric cereal cultivation on the North York Moors, England, *Journal of Archaeological Science*, 23, 613–18.

Singh, G., Joshi, R.D., Chopra, S.K. and Singh, A.B. (1974) Quaternary history of vegetation and climate of the Rajasthan desert, India, *Philosophical Transactions of the Royal Society of London, Series B*, 267, 467–501.

Singh, G., Kershaw, A.P. and Clark, R. (1981) Quaternary vegetation and fire history in Australia, in A.M. Gill, R.H. Groves and I.R. Noble (eds), *Fire and the Australian biota.* Canberra: Australian Academy of Science, pp. 23–54.

Sirocko, F., Sarnthein, M., Lange, H. and Erlenkeuser, H. (1991) Atmospheric summer circulation and coastal upwelling in the Arabian Sea during the Holocene and the last glaciation, *Quaternary Research*, 35, 72–93.

Sissons, J.B. (1980) Palaeoclimatic inferences from Loch Lomond Advance glaciers, in J.J. Lowe, J.M. Gray and J.E. Robinson (eds), *Studies in the Lateglacial of north-west Europe.* Oxford: Pergamon, pp. 31–43.

Smart, P.L. (1991) Uranium series dating, in P.L. Smart and P.D. Frances (eds), *Quaternary dating methods – a user's guide.* Technical Guide 4, Cambridge: Quaternary Research Association, pp. 45–83.

Smith, A.B. (1980) Domesticated cattle in the Sahara and their introduction into West Africa, in M.A.J. Williams and H. Faure (eds), *The Sahara and the Nile.* Rotterdam: A.A. Balkema, pp. 489–503.

Smith, A.G. (1970) The influence of Mesolithic and Neolithic man on British vegetation: a discussion, in D. Walker and R.G. West (eds), *Studies in the vegetational history of the British Isles: essays in honour of Harry Godwin.* Cambridge: Cambridge University Press, pp. 81–96.

Smith, A.G. (1981) The Neolithic, in I.G. Simmons and M.J. Tooley (eds), *The environment in British prehistory.* London: Duckworth, pp. 125–209.

Smith, A.G., Pearson, G.W. and Pilcher, J.R. (1971) Belfast radiocarbon dates III, *Radiocarbon*, 13, 103–25.

Smith, B.D. (1995) *The emergence of agriculture.* Freeman/Scientific American Library.

Smith, B.D. (1997) The initial domestication of *Cucurbita pepo* in the Americas 10 000 years ago, *Science*, 276, 932–4.

Smith, G.A. (1993) Missoula flood dynamics and magnitudes inferred from sedimentology of slack-water deposits on the Columbia plateau, Washington, *Bulletin of the Geological Society of America*, 105, 77–100.

Smith, C.I. and Street-Perrott, F.A. (1983) Pluvial lakes of the western United States, in S.C. Porter (ed.), *Late Quaternary environments of the United States. Volume 1, The Late Pleistocene*. London: Longman, pp. 190–212.

Smith, K., Coppen, J., Wainwright, G.J. and Beckett, S. (1981) The Shaugh Moor project: third report – settlement and environmental investigations, *Proceedings of the Prehistoric Society*, 47, 205–74.

Smol, J.P. (1986) Chrysophycean microfossils as indicators of lakewater pH, in J.P. Smol, R.W. Battarbee, R.B. Davis and J. Meriläinen (eds), *Diatoms and lake acidity*. Dordrecht: Dr W. Junk, pp. 275–87.

Smol, J.P. (1992) Paleolimnology: an important tool for effective ecosystem management, *Journal of Aquatic Ecosystem Health*, 1, 49–58.

Soudsky, B. and Pavlü, I. (1972) The Linear Pottery culture settlement patterns of central Europe, in P.J. Ucko, R. Tringham and G.W. Dimbleby (eds), *Man, settlement and urbanism*. London: Duckworth, pp. 317–28.

Spaulding, W.G., Leopold, E.B. and van Devender, T.R. (1983) Late Wisconsin paleoecology of the American Southwest, in S.C. Porter (ed.), *Late Quaternary environments of the United States. Volume 1, The Late Pleistocene*. London: Longman, pp. 259–93.

Stanley, D.J. and Warne, A.G. (1993) Sea level and initiation of Predynastic culture in the Nile delta, *Nature*, 363, 435–8.

Steadman, D.W., Stafford, T.W., Donahue, D.J. and Jull, A.J.T. (1991) Chronology of Holocene vertebrate extinction in the Galápagos Islands, *Quaternary Research*, 36, 126–33.

Stein, J.K. and Farrand, W.F. (eds) (1985) *Archaeological sediments in context*. Orono, Maine: Center for the Study of Early Man, University of Maine.

Stevenson, A.C. and Battarbee, R.W. (1991) Palaeoecological and documentary records of recent environmental changes in the Garaet el Ichkeul, *Biological Conservation*, 58, 275–95.

Stevenson, A.C. and Harrison, R.J. (1992) Ancient forests in Spain: a model for land-use and dry forest management in south-west Spain from 4000 BC to 1900 AD, *Proceedings of the Prehistoric Society*, 58, 227–47.

Steward, J. (1977) *Evolution and ecology*. Illinois: University of Illinois Press.

Stine, S. (1994) Extreme and persistent drought in California and Patagonia during mediaeval time, *Nature*, 369, 546–9.

Stokes, S., Thomas, D.S.G. and Washington, R. (1997) Multiple episodes of aridity in southern Africa since the last interglacial period, *Nature*, 388, 154–8.

Street, F.A. (1980) The relative importance of climate and local hydrogeological factors in influencing lake level fluctuations, *Palaeoecology of Africa*, 12, 137–58.

Street, F.A. and Grove, A.T. (1979) Global maps of lake-level fluctuations since 30 000 yr BP, *Quaternary Research*, 12, 83–118.

Street-Perrott, F.A., Huang, Y., Perrott, R.A., Eglinton, G., Barker, P., Ben Khelifa, L., Harkness, D.D. and Olago, D.O. (1997) Impact of lower atmospheric carbon dioxide on tropical mountain ecosystems, *Science* 278, 1422–6.

Street-Perrott, F.A. and Harrison, S.P. (1985) Lake levels and climate reconstruction, in A.D. Hecht (ed.), *Paleoclimate analysis and modeling*. Chichester: John Wiley, pp. 291–340.

Street-Perrott, F.A. and Perrott, R.A. (1990) Abrupt climatic fluctuations in the tropics: the influence of Atlantic Ocean circulation, *Nature*, 343, 607–12.

Street-Perrott, F.A. and Roberts, N. (1983) Fluctuations in closed basin lakes as an indicator of past atmospheric circulation patterns, in A. Street-Perrott, M. Beran and R. Ratcliffe (eds), *Variations in the global water budget*. Dordrecht: Reidel, pp. 331–45.

Street-Perrott, F.A., Marchand, D.S., Roberts, N. and Harrison, S.P. (1989) *Global lake-level variations from 18 000 to 0 years ago: a palaeoclimatic analysis*. Technical Report TR046, Washington, DC: US Department of Energy.

Street-Perrott, F.A., Mitchell, J.F.B., Marchand, D.S. and Brunner, J.S. (1991) Milankovitch and albedo forcing of the tropical monsoons: a comparison of geological evidence and numerical simulations for 9000 yr BP, *Transactions of the Royal Society of Edinburgh, Earth Science*, 81, 407–27.

Street-Perrott, F.A., Roberts, N. and Metcalfe, S. (1985) Geomorphic implications of late Quaternary hydrological and climatic changes in the northern hemisphere tropics, in I. Douglas and T. Spencer (eds), *Environmental change and tropical geomorphology*. London: George Allen and Unwin, pp. 165–83.

Stuiver, M. and Reimer, P.J. (1993) Extended ^{14}C data base and revised CALIB radiocarbon calibration program, *Radiocarbon*, 35, 215–30.

Sturm, M. (1979) Origin and composition of clastic varves, in C. Schlüchter (ed.), *Moraines and varves*. Rotterdam: Balkema, pp. 281–5.

Sutcliffe, D.W., Carrick, T.R., Heron, J., Rigg, E., Talling, J.F., Woof, C. and Lund, J.W.G. (1982) Long-term and seasonal changes in the chemical composition of precipitation and surface waters of lakes and tarns in the English Lake District, *Freshwater Biology*, 12, 451–506.

Swain, A.M. (1973) A history of fire and vegetation in northeastern Minnesota as recorded in lake sediments, *Quaternary Research*, 3, 383–96.

Talbot, M.R. (1990) A review of the palaeohydrological interpretation of the carbon and oxygen isotopic ratios in primary lacustrine carbonates, *Chemical Geology (Isotope Geosciences Section)*, 80, 261–79.

Talbot, M.R. and Delibrias, G. (1977) Holocene variations in the level of Lake Bosumtwi, Ghana, *Nature*, 268, 722–4.

Talbot, M.R., Livingstone, D.A., Palmer, P.G., Maley, J., Melack, J.M., Delibrias, G. and Gulliksen, S. (1984) Preliminary results from sediment core from Lake Bosumtwi, Ghana, *Palaeoecology of Africa*, 16, 173–92.

Tallis, J.H. and Switsur, V.R. (1990) Forest and moorland in the South Pennine uplands in the mid-Flandrian period. II. The hillslope forests, *Journal of Ecology*, 78, 857–83.

Tauber, H. (1965) Differential pollen dispersal and the interpretation of pollen diagrams, *Danmarks Geologiske Undersogelse II*, 89, 1–69.

Taylor, C. (1983) *Village and farmstead*. London: George Philip.

Taylor, D. (1990) Late Quaternary pollen records from two Ugandan mires: evidence for environmental change in the Rukiga Highlands of south-west Uganda, *Palaeogeography, Palaeoclimatology, Palaeoecology*, 80, 283–300.

Taylor, K.C., Hammer, C.U., Alley, R.B., Clausen, H.B., Dahl-Jenson, D., Gow, A.J., Gundestrup, N.S., Kipfstuhl, J., Moore, J.C. and Waddington, E.D. (1993) Electrical conductivity measurements from the GISP2 and GRIP Greenland ice cores, *Nature*, 366, 549–52.

Taylor, R.E., Stuiver, M. and Reimer, P.J. (1996) Development and extension of the calibration of the radiocarbon time scale: archaeological applications, *Quaternary Science Reviews (Quaternary Geochronology)*, 15, 655–68.

Teller, J.T. (1990) Volume and routing of late-glacial runoff from the southern Laurentide ice sheet, *Quaternary Research*, 34, 12–23.

Teller, J.T. and Last, W.M. (1990) Paleohydrological indicators in playas and salt lakes, with examples from Canada, Australia and Africa, *Palaeogeography, Palaeoclimatology, Palaeoecology*, 76, 215–40.

Thiébault, S. (1997) Holocene vegetation and human relationships in central Provence: charcoal analysis of the Baume de Fontbrégoua (Var, France), *The Holocene*, 7, 343–9.

Thomas, D.S.G. (1984) Ancient ergs of the former arid zones of Zimbabwe, Zambia and Angola, *Transactions of the Institute of British Geographers, N.S.*, 9, 75–88.

Thomas, K.D. (1985) Land snail analysis in archaeology: theory and practice, in N.R.J. Fieller, D.D. Gilbertson and N.G.A. Ralph (eds), *Palaeobiological investigations. Research design, methods and data analysis*. Oxford: British Archaeological Reports, International Series 266, pp. 131–56.

Thomas, M.F. and Thorp, M.B. (1995) Geomorphic response to rapid climatic and hydrologic change during the late Pleistocene and early Holocene in the humid and sub-humid tropics, *Quaternary Science Reviews*, 14, 101–24.

Thompson, L.G. (1991) Ice-core records of with emphasis on the global record of the last 2000 years, in R.S. Bradley (ed.), *Global changes of the past*. Boulder, Colorado: OIES Global Change Institute, pp. 201–24.

Thompson, L.G., Mosley-Thompson, E., Dansgaard, W. and Grootes, P.M. (1986) The 'Little Ice Age' as recorded in the stratigraphy of the tropical Quelccaya ice cap, *Science*, 234, 361–4.

Thompson, L.G., Mosley-Thompson, E., Davis, M.E., Lin, P.-N., Henderson, K.A., Cole-Dai, J., Bolzan, J.F. and Liu, K.-B. (1995) Late glacial stage and Holocene tropical ice core records from Huascarán, Peru, *Science*, 269, 46–50.

Thompson, R. (1984) A global review of palaeomagnetic results from wet lake sediments, in E.Y. Haworth and J.W.G. Lund (eds), *Lake sediments and environmental history*. Leicester: Leicester University Press, pp. 145–64.

Thompson, R. and Oldfield, F. (1986) *Environmental magnetism*. London: George Allen and Unwin.

Thordarson, T.H. and Self, S. (1993) The Laki (Skaftár Fires) and Grímsvötn eruptions in 1783–1785, *Bulletin of Volcanology*, 55, 233–63.

Thorne, A. (1980) The arrival of man in Australia, in A. Sherratt (ed.), *Cambridge Encyclopaedia of Archaeology*. Cambridge: Cambridge University Press, pp. 96–100.

Thornes, J.B. (1987) The palaeo-ecology of erosion, in J.M. Wagstaff (ed.), *Landscape and culture*. Oxford: Blackwell, pp. 37–55.

Tipping, R.M. (1985) Loch Lomond stadial *Artemisia* pollen assemblages and Loch Lomond readvance regional firn-line altitudes, *Quaternary Newsletter*, 46, 1–11.

Tooley, M.J. and Shennan, I. (eds) (1987) *Sea-level changes*. Oxford: Blackwell.

Tricart, J. (1972) *Landforms of the humid tropics, forests and savannas*. London: Longman.

Tricart, J. (1974) Existence de périodes sèches au Quaternaire en Amazonie et dans les régions voisines, *Revue de Géomorphologie Dynamique*, 23, 145–58.

Troels-Smith, J. (1955) Karakterisering af løse jordarter (characterization of unconsolidated sediments), *Danmarks Geologiske Undersøgelse*, IV (3), 1–73.

Troels-Smith, J. (1960) Ivy, mistletoe and elm: climatic indicators – fodder plants, *Danmarks Geologiske Undersøgelse*, IV (4), 1–32.

Tsukada, M. (1983) Vegetation and climate during the last glacial maximum in Japan, *Quaternary Research*, 19, 212–35.

Turner, C. (1970) The middle Pleistocene deposits at Marks Tey, Essex, *Philosophical Transactions of the Royal Society of London, Series B*, 257, 373–440.

Turner, C. and Hannon, G.E. (1988) Vegetational evidence for late Quaternary climatic changes in southwest Europe in relation to the influence of the North Atlantic Ocean, *Philosophical Transactions of the Royal Society of London, Series B*, 318, 451–85.

Turner, J. (1964) Anthropogenic factor in vegetation history, *New Phytologist*, 63, 73–89.

Turner, J. (1975) The evidence for land use by prehistoric farming communities: the use of three-dimensional pollen diagrams, in J.G. Evans, S. Limbrey and H. Cleere (eds), *The effect of man on the landscape: the highland zone*. C.B.A. Research Report 11, pp. 86–95.

Turner, J. (1981) The Iron Age, in I. Simmons and M. Tooley (eds), *The environment in British prehistory*. London: Duckworth, pp. 250–81.

Turner, R.E. and Rabalais, N.N. (1994) Coastal eutrophication near the Mississippi river delta, *Nature*, 368, 619–21.

Tzedakis, P.C. (1993) Long-term tree populations in northwest Greece through multiple Quaternary climatic cycles, *Nature*, 364, 437–40.

Tzedakis, P.C., Bennett, K.D. and Magri, D. (1994) Climate and the pollen record, *Nature*, 370, 513.

Vandenberghe, J. and Bohncke, S. (1985) The Weichselian late glacial in a small lowland valley (Mark river, Belgium and the Netherlands), *Bulletin de l'Association française pour l'étude du Quaternaire*, 167–75.

Vandenberghe, J., Paris, P., Kasse, C., Gouman, M. and Beyens, L. (1984) Paleomorphological and botanical evolution of small lowland valleys, *Catena*, 11, 229–38.

Vartanyan, S.L., Garutt, V.E. and Sher, A.V. (1993) Holocene dwarf mammoths from Wrangel Island in the Siberian Arctic, *Nature*, 362, 337–40.

Vereshschagin, N.K. and Baryshnikov, G.F. (1984) Quaternary mammalian extinctions in northern Eurasia, in P.S. Martin and R.G. Klein (eds), *Quaternary extinctions*. Tucson: University of Arizona Press, pp. 483–516.

Vernet, J.-L. and Thiébault, S. (1987) An approach to northwestern Mediterranean recent prehistoric vegetation and ecologic implications, *Journal of Biogeography*, 14, 117–27.

Viner, A.B. and Smith, I.R. (1973) Geographical, historical and physical aspects of Lake George, *Proceedings of the Royal Society of London, Series B*, 184, 235–70.

Vita-Finzi, C. (1969a) Geological opportunism, in P.J. Ucko and G.W. Dimbleby (eds), *The domestication and exploitation of plants and animals*. London: Duckworth, pp. 31–4.

Vita-Finzi, C. (1969b) *The Mediterranean valleys*. Cambridge: Cambridge University Press.

Vita-Finzi, C. (1973) *Recent earth history*. London: Macmillan.

Vita-Finzi, C. (1974) Related territories and alluvial sediments, in E.S. Higgs (ed.), *Palaeoeconomy*. Cambridge: Cambridge University Press, pp. 225–31.

Vita-Finzi, C. (1978) *Archaeological sites in their setting*. London: Thames and Hudson.

Vita-Finzi, C. and Roberts, N. (1984) Selective leaching of shells for ^{14}C dating, *Radiocarbon*, 26, 54–8.

Wagstaff, J.M. (1981) Buried assumptions: some problems in the interpretation of the 'Younger Fill' raised by recent data from Greece, *Journal of Archaeological Science*, 8, 247–64.

Walcott, R.I. (1972) Past sea levels, eustasy and deformation of the Earth, *Quaternary Research*, 2, 1–14.

Walker, D. (1970) Direction and rate of change of some British postglacial hydroseres, in D. Walker and R.G. West (eds), *Studies in the vegetation history of the British Isles*. Cambridge: Cambridge University Press, pp. 117–39.

Walker, I.R., Mott, R.J. and Smol, J.P. (1991a) Allerød-Younger Dryas lake temperatures from midge fossils in Atlantic Canada, *Science*, 253, 1010–12.

Walker, I.R., Smol, J.P., Engstrom, D.E. and Birks, H.J.B. (1991b) An assessment of Chironomidae as quantitative indicators of past climatic change, *Canadian Journal of Fisheries and Aquatic Sciences*, 48, 975–87.

Wallerstein, I. (1974) *The modern world-system: capitalist agriculture and the origins of the European world economy in the sixteenth century*. London: Academic Press.

Walling, D.E. and Quine, T.A. (1990) Use of Caesium-137 to investigate patterns and rates of soil erosion on arable fields, in J. Boardman, I.D.L. Foster and J.A. Dearing (eds), *Soil erosion on agricultural land*. Chichester: John Wiley, pp. 33–53.

Warner, B.G. (1988) Methods in Quaternary ecology 5. Testate amoebae (Protozoa), *Geoscience Canada*, 15, 251–60.

Waton, P.V. (1982) Man's impact on the chalklands: some new pollen evidence, in M. Bell and S. Limbrey (eds), *Archaeological aspects of woodland ecology*. Oxford: British Archaeological Reports, International Series 146, pp. 75–92.

Watson, R.A. and Wright, Jr, H.E. (1980) The end of the Pleistocene: a general critique of chronostratigraphic classification, *Boreas*, 9, 153–63.

Watts, W.A. (1980) Regional variation in the response of vegetation to late glacial climatic shifts, in J.J. Lowe, J.M. Gray and J.E. Robinson (eds), *Studies in the Lateglacial of north-west Europe*. Oxford: Pergamon, pp. 1–22.

Watts, W.A. (1983) Vegetational history of the eastern United States 25 000 to 10 000 years ago, in S.C. Porter (ed.), *Late-Quaternary environments of the United States. Volume 1, The Late Pleistocene*. London: Longman, pp. 294–310.

Watts, W.A. (1984) The Holocene vegetation of the Burren, western Ireland, in E.Y. Haworth and J.W.D. Lund (eds), *Lake sediments and environmental history*. Leicester: Leicester University Press, pp. 359–76.

Watts, W.A., Allen, J.R.M., Huntley, B. and Fritz, S.C. (1996) Vegetation history and climate of the last 15 000 years at Laghi de Monticchio, southern Italy, *Quaternary Science Reviews*, 15, 113–32.

Webb, R.S. and Webb, T. (1988) Rates of accumulation in pollen cores from small lakes and mires of eastern North America, *Quaternary Research*, 30, 284–97.

Webb, T. (1985) Holocene palynology and climate, in A.D. Hecht (ed.), *Paleoclimate analysis and modeling*. Chichester: John Wiley, pp. 163–95.

Webb, T. (1986) Is the vegetation in equilibrium with climate? How to interpret late-Quaternary pollen data, *Vegetation*, 67, 75–91.

Webb, T., Bartlein, P.J., Harrison, S.P. and Anderson, K.H. (1993) Vegetation, lake levels and climate in eastern North America for the past 13 000 years, in H.E. Wright, Jr, J.E. Kutzbach, T. Webb III, W.F. Ruddiman, F.A. Street-Perrott and P.J. Bartlein (eds), *Global climates since the last glacial maximum*. Minneapolis: University of Minnesota Press, pp. 415–67.

Webb, T., Cushing, E.J. and Wright, H.E. (1984) Holocene changes in the vegetation of the Midwest, in H.E. Wright (ed.), *Late Quaternary environments of the United States. Volume 2, The Holocene*. London: Longman, pp. 142–65.

Weiss, H., Courty, M.A., Wetterstrom, W., Guichard, F., Senior, L., Meadow, R. and Curnow, A. (1993) The genesis and collapse of third millennium North Mesopotamian civilization, *Science*, 261, 995–1004.

Wells, G.L. (1983) Late-glacial circulation over central North America revealed by aeolian features, in A. Street-Perrott, M. Beran and R. Ratcliffe (eds), *Variations in the global water budget*. Dordrecht: Reidel, pp. 317–30.

Wells, P.V. (1976) Macrofossil analysis of woodrat (*Neotoma*) middens as a key to the Quaternary vegetational history of arid America, *Quaternary Research*, 6, 223–48.

Wendorf, F., Schild, R., Said, R., Haynes, C.V., Gautier, A. and Kouseiwicz, M. (1976) The prehistory of the Egyptian Sahara, *Science*, 193, 103–14.

West, F.H. (1983) The antiquity of man in America, in S.C. Porter (ed.), *Late Quaternary environments of the United States. Volume 1, The Late Pleistocene*. London: Longman, pp. 364–82.

Whitney, G.G. (1994) *From coastal wilderness to fruited plain: a history of environmental change in temperate North America from 1500 to the present*. Cambridge: Cambridge University Press.

Whittle, A.W.R. (1978) Resources and population in the British Neolithic, *Antiquity*, 52, 34–42.

Wijmstra, T.A. (1969) Palynology of the first 30 m of a 120 m deep section in northern Greece, *Acta Botanica Neerlandica*, 18, 511–27.

Wilkes, G. (1989) Maize: domestication, racial evolution, and spread, in D.R. Harris and G.C. Hillman (eds), *Farming and foraging: the evolution of plant exploitation*. London: Unwin Hyman, pp. 440–5.

Wilkinson, T.J. (1982) The definition of ancient manured zones by means of extensive sherd-sampling techniques, *Journal of Field Archaeology*, 9, 323–33.

Wilkinson, T.J. (1989) Extensive sherd scatters and land-use intensity: some recent results, *Journal of Field Archaeology*, 16, 31–46.

Williams, D.F., Thunell, R.C. and Kennett, J.P. (1978) Periodic freshwater flooding and stagnation of the eastern Mediterranean Sea during the late Quaternary, *Science*, 201, 252–4.

Williams, M.A.J., Dunkerley, D.L., de Deckker, P., Kershaw, A.P. and Stokes, T. (1993) *Quaternary environments*. London: Arnold.

Williams, N.E. (1988) The use of caddisflies (Trichoptera) in palaeoecology, *Palaeogeography, Palaeoclimatology, Palaeoecology*, 62, 493–500.

Willis, K.J. and Bennett, K.D. (1994) The Neolithic transition: fact or fiction? Palaeoecological evidence from the Balkans, *The Holocene*, 4, 326–30.

Willis, K.J., Braun, M., Sümegi, P. and Tóth, A. (1997) Does soil change cause vegetation change or vice versa? A temporal perspective from Hungary, *Ecology*, 78, 740–50.

Wintle, A.G. and Huntley, D.J. (1982) Thermoluminescence dating of sediments, *Quaternary Science Reviews*, 1, 31–53.

Wise, S.M. (1980) Caesium-137 and lead-210: a review of the techniques and some applications in geomorphology, in R.A. Cullingford, D.A. Davidson and J. Lewin (eds), *Timescales in geomorphology*. Chichester: John Wiley, pp. 109–27.

Wise, S.M., Thornes, J.B. and Gilman, A. (1982) How old are the badlands? A case-study from south-east Spain, in R. Bryan and A. Yair (eds), *Badland geomorphology and piping*. Norwich: Geobooks, pp. 259–78.

Wittfogel, K.A. (1956) The hydraulic civilizations, in W.L. Thomas (ed.), *Man's role in changing the face of the earth*. Chicago: University of Chicago Press, pp. 152–64.

Wittfogel, K.A. (1957) *Oriental despotism: a comparative study of total power*. New Haven: Yale University Press.

Woillard, G. (1979) Abrupt end of the last interglacial s.s. in northeast France, *Nature*, 281, 558–62.

Wolman, M.G. (1967) A cycle of sedimentation and erosion in urban river channels, *Geografiska Annaler*, 49A, 385–95.

Woodman, P.C. (1978) The chronology and economy of the Irish Mesolithic: some working hypotheses, in P. Mellars (ed.), *The early postglacial settlement of northern Europe*. London: Duckworth, pp. 333–70.

Woodman, P.C. (1985) *Excavations at Mount Sandel, 1973–77*. Belfast: HMSO.

Wright, H.E. (1976) The environmental setting for plant domestication in the Near East, *Science*, 194, 385–9.

Wright, H.E. and Heinselman, M.L. (1973) The ecological role of fire in natural conifer forests of western and northern North America – introduction, *Quaternary Research*, 3, 319–28.

Wright, Jr, H.E. (1993) Environmental determinism in Near Eastern prehistory, *Current Anthropology*, 34, 458–69.

Wright, Jr, H.E., Kutzbach, J.E., Webb III, T., Ruddiman, W.F., Street-Perrott, F.A. and Bartlein, P.J. (eds) (1993) *Global climates since the last glacial maximum*. Minneapolis: University of Minnesota Press.

Wright, R.F. (1977) *Historical changes in the pH of 128 lakes in southern Norway and 130 lakes in southern Sweden over the period 1923–1976*. SNSF project TN 34/77.

van Zeist, W. and Bottema, S. (1977) Palynological investigations in western Iran, *Palaeohistoria*, 19, 19–85.

van Zeist, W., Woldring, H. and Stapert, D. (1975) Late Quaternary vegetation and climate of southwestern Turkey, *Palaeohistoria*, 17, 53–143.

van Zeist, W. and Bottema, S. (1991) *Late Quaternary vegetation of the Near East*. Beihefte zum Tübinger Atlas des Vorderen Orients, Reihe A18. Wiesbaden: Dr L. Reichert Verlag, 156 pp.

Zielinski, G.A., Mayewski, P.A., Meeker, L.D., Whitlow, S., Twickler, M.S., Morrison, M., Meese, D.A., Gow, A.J. and Alley, R.B. (1994) Record of volcanism since 7000 BC from the GISP2 Greenland ice core and implications for the volcano-climate system, *Science*, 262, 948–52.

Zohary, D. (1989) Domestication of the Southwest Asian Neolithic crop assemblage of cereals, pulses and flax: the evidence from the living plants, in D.R. Harris and G.C. Hillman (eds), *Farming and foraging: the evolution of plant exploitation*. London: Unwin Hyman, pp. 335–43.

Zohary, D. and Hopf, M. (1993) *Domestication of plants in the Old World. The origin and spread of cultivated plants in west Asia, Europe and the Nile valley*, 2nd edn. Oxford: Clarendon Press.

Zohary, D. and Spiegel-Roy, P. (1975) Beginnings of fruit growing in the Old World, *Science*, 187, 319–27.

Zvelebil, M. (1980) The rise of the nomads in central Asia, in A. Sherratt (ed.), *The Cambridge Encyclopaedia of Archaeology*. Cambridge: Cambridge University Press, pp. 252–6.

Zvelebil, M. (ed.) (1986) *Hunters in transition*. Cambridge: Cambridge University Press.

INDEX